THE
WORLD OF
CARS

THE WORLD OF CARS

ROY BACON

With a Foreword
by
LORD MONTAGU OF BEAULIEU

SUNBURST BOOKS

This edition first published in 1995 by
Sunburst Books,
Deacon House,
65 Old Church Street
London SW3 5BS

ISBN 1 85778 206 2

Printed in Hong Kong

CONTENTS

FOREWORD

My father, a pioneer motorist since 1898 and the first parliamentary champion of the motorist's cause, created the link between the Montagu family and motoring. Over the years, that link has strengthened into what it is today, and has come to mean much to both myself and Beaulieu.

In 1952, I opened Palace House and its gardens to the public, displaying a handful of early vehicles in the Front Hall of the house. From this modest start the Museum began to grow, or rather evolve, as various items and vehicles were added to the displays within Palace House. By 1958, with Palace House being taken over by cars and beginning to smell a little like a garage, it became clear the exhibits would have to be housed in a purpose-built building within the grounds.

In 1959 the collection officially became recognised as a Motor Museum and was moved to new premises in the grounds, but even this was to prove inadequate. A long-term plan was drawn up in the mid 1960s, intended to increase accommodation for the museum and improve tourist facilities while retaining a harmonious aspect with the countryside and the historical buildings of Beaulieu. A new museum, restaurant and car parks were to be fitted into the landscape so as to be virtually invisible from outside the estate. The leading architects called in to undertake this did their job well and the result is an astonishing success. At the same time a charitable trust was established to own the collection and in 1972 the collection was reopened as the National Motor Museum.

In the National Motor Museum, I have brought together many of the great classic cars of the past hundred years, including several of my own particular dream cars. The definition of what exactly is a dream car is always subjective, of course, and I would not imagine that my own particular taste matches that of everyone who reads this book. Indeed, there are probably as many dream cars as there are dreamers.

Some prefer the golden age of American motoring: the great finned and chromed motor cars of the United States in the 1950s. Such dreams are made of long, highly stylised bodies, powerful lazy engines and the way of life that went hand-in-hand with these cars. Others dream of speed and performance – the pursuit of the ultimate race-bred thoroughbred. Fast cars, however, lose much of their glamour and importance when each new, faster model pushes its predecessor into the realm of forgotten dreams, never to be satisfied and always to be chased.

The cars of elegance and luxury that offer every conceivable creature comfort and an unsurpassed opulence are certainly objects of many a desire. The famed marques of Rolls-Royce, Cadillac and Bugatti give a ride so smooth and serene that it would seem the car is transported on an air cushion rather than wheels and mechanics. Or perhaps the great dream cars are those which have the elusive and indefinable quality known as 'head-turning ability' – the cars endowed with a spark of genius, combining grace and beauty with superb engineering.

To mention every dream car of the past, those that enthusiasts love to discuss whenever they gather, would be impossible. But some certainly stand head and shoulders – or grill and bonnet – above the others. There is the great Bentley Blower, more properly termed the 4½-litre Supercharged Bentley, the Alpha-Romeo 8C 2300, the 540K Mercedes Benz and the type 37 Bugatti; all grand sports cars which lay claim to legendary greatness. Rolls-Royce motor cars, both vintage and modern, epitomise the taste, elegance and status of luxury cars, while the Hispano-Suizas embody the grace and romance of Continental motoring before the First World War.

Later cars also are among the most desirable. Lamborghinis and Jaguars rank among the best ever built. And the proliferation of less ostentatious sports cars has allowed those with more modest pockets to sample the delights of MG, Triumph and Fiat, but the ultimate dreams have to be the supercars of recent years – the Jaguar XJ220, McLaren F1 and the Ferrari F40.

Today, whatever the dream, there is a car to fit. Since that first horseless carriage rattled down the street and began the irreversible progress of the automobile, people have dreamed of cars. And cars have been the dreams of people.

Montagu of Beaulieu

INTRODUCTION

The car has been with us for over a century to stir our emotions, fire our passions, bring joy and sorrow, and transport us around. There are now few places on earth where wheel tracks have not passed by, always leaving much more than a tyre imprint in the ground. Most remote is the second-hand vehicle marooned on the moon, just one careful owner.

In the early days of the primitives, completing a journey was an achievement in itself. In many countries drivers had to contend with restrictions, obstructions and legal hounding, quite apart from any troubles relating to the mechanics of the car. Vested interests in coal mining, trains, horses and trams raised all manner of obstacles to the automobile and some, it seems, have never really gone away.

Now, we step into our carriage to be surrounded by high technology that touches ever facet of the motor car and its driver. Speed, handling, brakes, economy, safety, comfort, all move on relentlessly, each decade reaching new heights that seemed impossible in the one that preceded it.

Men and women love their motorcars. Despite the frequent pressures from politicians, bureaucrats and environmentalists to reduce or prevent the use of cars, their numbers are ever on the increase. The basic idea of personal transport from door to door is as old as time the world over, using animals to ride on or to pull a chariot.

A century ago the lord had his coach, the squire his carriage and the vicar his trap, while most others had to walk locally or to the station for the train on the rare occasion they travelled any distance. Now, we can all share the dream and the reality with its benefits. Children to school, the ride to work, shopping, visiting places and friends, all are now available and desirable.

Hardly surprising that for most of us the car is our second most-valued possession and for a good number it takes pride of place in our hearts and minds while house and home go untouched. It is truly the vehicle for our dreams, some personal, some family, some possible, some to stay simply dreams but nonetheless worth having.

Here is no history book, but a pictorial look at the cars that have stirred people over the last century. If every picture is worth a thousand words then here is a fat tale indeed, but more of reminiscence than a story. A place to trigger memories, of cars owned, cars seen, cars that were trouble, cars of joy, cars on journeys both magical and dreadful, cars that stay in the mind forever.

Everyone will see each picture differently, and every sentence that begins 'Do you remember when...' will have its own ending. What will yours be?

AC

This firm began in Edwardian days, building a three-wheeled Autocarrier, progressed to similar passenger machines and on to four wheels. In the 1920s they introduced a fine in-line, six-cylinder, overhead-camshaft, 2-litre engine that they were to use right up to the 1960s. During the 1920s AC built fine cars in vintage style, while sporting ones predominated during the 1930s. Postwar, they built a three-wheeler and a saloon before turning to sports cars.

1960 AC GREYHOUND
Derived from the sporty AC Ace, the Greyhound was a four-seat saloon, powered by a 2-litre, triple-carburettor, Bristol-built six-cylinder engine which replaced the old AC engine introduced during the 1920s. There was independent suspension all round, front disc brakes and an optional overdrive for the four-speed gearbox to complete the specification.

1965 AC SHELBY COBRA
The firm began to build the two-seater Ace sports car in 1953 using a tubular chassis, independent suspension all round and the old AC-six engine. Later, Carroll Shelby proposed the Cobra, fitting a 4.2-litre Ford V-8 engine in a strengthened chassis for fast road or, as here, competition use.

1965 AC SHELBY COBRA The stark, competition cockpit of a model which was successful at Le Mans.

1966 AC COBRA 427 The Cobra moved on to a 7-litre Ford engine in 1964 to become a leading muscle-car with 400bhp under the hood, sometimes even more in street-legal variants. Not for the faint-hearted. The side exhaust (below) was in the best Cobra traditions but was hardly the quietest on the street.

1981 AC 3000ME Production of the tough Cobra stopped in 1973 but it was six years before this replacement finally went into production. The 3-litre V-6 Ford engine was mounted in a two-seat coupé body transversely behind the seats. The handling and performance were not up to expectations. Few were made and production ceased in 1984.

1990 AC COBRA AUTOKRAFT Autokraft and Ford became equal partners in AC, owning both the company and Cobra names. Manufacture of the model began again in 1983, keeping to the format as of old, and continues for those who seek the same combination of power, speed and handling as in the past.

1990 DAX COBRA TOJEIRO The use of John Tojeiro's name recalls his work on the original Ace chassis design, the model that led to the Cobra.

ALFA ROMEO

One of Italy's best-known and finest firms that has built sports and sporting cars since 1910. Founded as Anonima Lombarda Fabbrica Automobili that year, Nicolo Romeo took control in 1915 and the 1920s brought fine designs from Giuseppe Merosi and Vittorio Jano. The firm had a most successful racing team in the 1930s, run by Enzo Ferrari, and made many fine sporting models prewar. Postwar, Alfa Romeo concentrated more on saloons, but the marque experienced many problems during the 1970s. Fiat took it over in 1987, and the new management's efforts have made Alfa Romeo more successful.

1929 ALFA ROMEO 6C 1500
One of the classic designs from Vittorio Jano, derived from the firm's racing engines, had a 1.5-litre, six-cylinder engine with twin overhead camshafts. This went into a fine chassis to combine both excellent handling and brakes in a superb car.

1932 ALFA ROMEO 8C 2300
Again a classic, in this case with a 2.3-litre, straight-eight engine whose two cylinder blocks had the drive to the twin overhead camshafts between them. It was essentially a detuned racing engine with dry-sump lubrication in a car well able to exceed 160 km/100 miles per hour.

**1938 & 1939 ALFA ROMEO
8C 2900B** As with the 8C 2300, these two very sporting road cars also used a modified racing engine. Of 2.9 litres, it came from the P3 works racer complete with twin camshafts, twin superchargers and magneto ignition. There was at least 180bhp for the multi-plate clutch and four-speed gearbox transmit. All wheels had independent suspension. Both the coupé and the open body (below) had great style while the works Superleggera version won the 1938 Mille Miglia.

ABOVE: **1948 ALFA ROMEO 6C 2500** This 2.4-litre model was introduced late in 1939 but production continued after the war until 1953, still with the lovely twin-cam engine. In this case it was fitted with a fine, elegant open body.

LEFT: **1955 ALFA ROMEO BAT 9** Leading body stylist Nuccio Bertone built a series of show cars for Alfa Romeo, and this is the last. This type of exercise was and still is common practice, producing concept cars for shows to indicate future directions in style and design.

ABOVE: **1974 ALFA ROMEO 1600 SPIDER** From the late 1950s many open models appeared, based on existing closed bodies and engines. Pininfarina produced the Spider bodies from 1966 and they were at first known as the Duetto.

BELOW: **1975 ALFA ROMEO GTV 2000 SPIDER** A further step in the open-body line, using the 1962cc version of the twin-cam engine whose 132bhp output would run the car up to 190 km/118 miles per hour. It continued in production into the 1990s.

1975 ALFA ROMEO 1600GTV JUNIOR Many Alfas of the 1970s combined existing engines and bodies to extend the range. This coupé is a Veloce fitted with the 1570cc twin-cam engine.

1981 ALFA ROMEO SPRINT VELOCE The Alfasud was launched in 1972 fitted with an 1186cc, flat-four, overhead-camshaft engine driving the front wheels. Its capacity soon grew to take advantage of the great handling and it was quickly joined by the TI and Sprint versions. This one has a 1.5-litre engine, and an elegant coupé body from the ItalDesign studio.

1984 ALFA ROMEO 2.0 GTV This performance coupé was launched in 1976 using the 1962cc, twin-cam, four-cylinder engine in an ItalDesign body and was able to run to 190 km/118 miles per hour. Later, it was given a 2492cc, overhead-camshaft, V-6 engine which could push it to 210 km/130 miles per hour.

ABOVE: **1989 ALFA ROMEO 75** The big Alfa Romeo saloon for the late 1980s combined fine handling with a choice of engines ranging from 1.8 to 3 litres in four-cylinder or V-6 formats. This has the 2-litre Twin Spark four with dual plugs per cylinder. A rear-mounted gearbox aided balance if not changing gears.

BELOW: **1991 ALFA ROMEO 2000 SPIDER** The sports coupé format which had been used for so long retained some of the 1966 Pininfarina line while using the 2-litre, twin-cam engine for performance.

ALLARD

Before the war Sydney Allard was a successful trials driver and began to build similar cars for others based on Ford parts. Postwar, he went into production with this stylish sports car (below, a 1948 model) still using a Ford V-8 engine, offering outstanding performance. Later models used larger American V-8s up to 1959.

ALVIS

Founded in 1919, this great British marque was known in the 1920s for its 12/50 sports car and its use of front-wheel drive. In the next decade the accent was on luxury, resulting in some fine saloons with stylish bodies. Postwar models were less inspiring and car production finally ceased in 1967, two years after Rover acquired the firm.

1928 ALVIS FWD Much in the style of their other models of the decade, this front-wheel-drive sports model with a two-seater body had a 1.5-litre, overhead-camshaft engine plus the option of a supercharger. Although the firm had demonstrated the car's abilities at Le Mans, the public lacked confidence in the technology, so sales suffered.

RIGHT: **1937 ALVIS SPEED 25**
The Speed models of the 1930s came in a variety of body types – sports tourer, coupé, sports saloon and custom built. The original 2.5-litre, six-cylinder engine was enlarged to 3.6 litres for this model which was well appointed and had the fine lines of the later part of the decade.

1937 ALVIS SPEED 25 SEDANCA The fine lines of this Sedanca body give another kind of 1930s elegance to the Speed 25.

1950 ALVIS TB21 First seen in 1948 as the TB14, this two-seater sports car adopted the curvaceous postwar style for the body. At first it fitted the 1.9-litre, four-cylinder engine but then changed to a 3-litre six. It carries the famous red triangular radiator badge.

1953 ALVIS TA21 An elegant car, based on the earlier TA14 but with the 3-litre, six-cylinder engine, independent front suspension and saloon or drophead coupé bodies.

1963 ALVIS TD21 A later drophead coupé which, with a similar saloon, took the firm up to 1967 when car production ceased. They were fine cars, but expensive.

AMPHICAR

The notion of a car able to be used as a boat lives more in James Bond films but has occurred. This German design (below, 1965 Amphicar), created by Hans Trippel who had experience of such vehicles for the services, was built in Europe, powered by an English Triumph engine and sold mainly in the United States.

ARMSTRONG SIDDELEY

Formed in 1919 by an aircraft manufacturer, this firm built quality cars from the start, offering comfort and ease of control. This continued between the wars and after. The marque was perceived as well made, although postwar models were too expensive for their traditional market. Car manufacture ceased in 1960.

1935 ARMSTRONG SIDDELEY During the 1930s the company built its cars with a variety of body styles, various sizes of six-cylinder engine, and alternative wheelbases. This is one of the many from that period when engines reached 5 litres for the limousines.

1949 ARMSTRONG SIDDELEY HURRICANE Postwar models reflected the firm's aviation connections, hence this Hurricane drophead coupé and the Lancaster saloon which fitted a 2 or 2.3-litre, six-cylinder engine. The sphinx mascot (above) adorned the cars throughout production, the motto – 'as silent as the sphinx' – part of the image.

ARNOLD

Based in East Peckham, London, Arnold imported Benz cars before building a dozen under their own name. The result (below, an 1896 Arnold) was powered by an 1190cc, single-cylinder engine. While they may have been the first British firm to embark on series production, the venture was not a success.

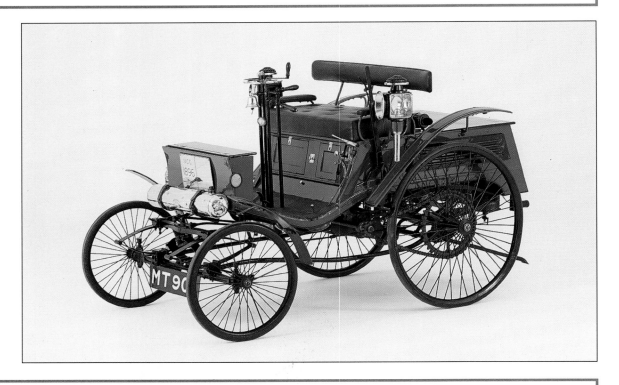

ARROL JOHNSTON

A Scottish firm, based at first in Paisley, who built this dog-cart (below, 1902) powered by an opposed-twin engine with four pistons for some 10 years. Their 1905 model had an improved appearance but the same form of engine, and was followed by other variants. In 1919 their overhead-camshaft Victory model was under-developed and hastily withdrawn, to be replaced by a prewar one that saw them through to their 1929 end.

ASTON MARTIN

Their car production began in 1921 with a sports model and since then they have always built sporting open and saloon cars. Through the 1930s they tended to put on weight, but without losing their style, and in 1947 they were taken over by David Brown. Then came the fabulous DB-series, a high profile racing programme, fame in the James Bond films and, eventually, control by Ford.

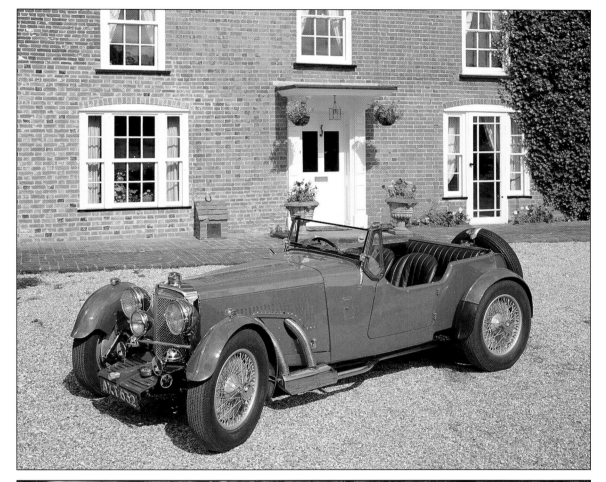

1934 ASTON MARTIN LE MANS Derived from the earlier International model, the Le Mans continued with the 1.5-litre, four-cylinder engine with chain-driven overhead camshaft. Cycle wings, outside exhaust and slab petrol tank were all marks of the prewar sports car.

1952 ASTON MARTIN DB2 A twin overhead camshaft, six-cylinder engine designed by W. O. Bentley, tubular space frame and a wonderful body style made up the DB2, the start of a long and successful line. The interior (above) was hardly ergonomic but was typical of the time and derived from a prewar Lagonda.

ABOVE: **1958 ASTON MARTIN DB MK III** The next stage in the line brought more power from the engine, Girling disc brakes, a new and smoother grill and the fine treatment of the rear body as it developed. The instruments were regrouped in front of the driver and the result one of the best-looking coupés of the earlier cars built after the 1948 move to Feltham.
RIGHT: **1965 ASTON MARTIN DB5** The car made famous by the Bond films, complete with its rotating number plates, tyre shredders, oil spray, passenger ejection seat, machine guns and bulletproof rear panel. Inside went the radar screen for tracking villains. A rear view (below) shows two of the devices

1975 ASTON MARTIN LAGONDA David Brown took over Lagonda in 1947 and revived the name for this model based on the Aston engine and mechanics. It had unusual, sharp-edged styling and, at first, complex electronics and instrumentation soon simplified to eliminate many problems. Its style did not last long but its opulence kept it selling to 1990.

1987 ASTON MARTIN VOLANTE This four-seater convertible was based on the Vantage saloon, both of which used the 5340cc, twin-cam, V-8 engine. This produced 305bhp in this application while the body had extra stiffening and a power hood. Not quite as quick as the saloon but more than adequate for the United States and Middle East where style counted.

ABOVE: **1992 ASTON MARTIN VIRAGE** A very fine and very expensive motor car using the same V-8 engine as the Volante, but hand-built and given four-valve heads which produced 310bhp and ran to 250 km /155 miles per hour. The body was crafted in aluminium in a most elegant form that looked smaller than it was.

RIGHT: **1994 ASTON MARTIN VANTAGE** The Vantage name was first used to indicate an engine option but later became a model developed from the DBS. This continued with the cars fitted with the V-8 engine where the car not only had extra power but also a number of revisions to the chassis and body.

AUBURN

This firm was located in Auburn, Indiana, and built its first car in 1900, moving into production in 1903. In 1925 they introduced a fine straight-eight engine and the Speedster model. Success followed but they were hit hard by the Depression although they managed to keep going up to 1937.

1932 AUBURN V-12 SPEED-STER Despite the Depression, the firm added models fitted with a 6390cc, V-12 engine in a stretched chassis having servo brakes and the option of a two-speed rear axle. In this case it had a classic convertible body with the pen-nib two-tone bonnet finish.

1933 AUBURN BOAT-TAIL SPEEDSTER The models with the straight-eight engine continued alongside the V-12 and in the same variety of bodies. This boat-tailed one is handsome and a classic.

ABOVE: **1933 AUBURN 8-105** A further variation, typical of the early 1930s, to the body highlights the lines and the marque. The pen-nib line extends from the bonnet and along the doors to set the car off so well.

BELOW: **1935 AUBURN 851S** For their final model, the firm added a supercharger to the 4596cc straight-eight, side-valve engine which raised the power further. There were plated exhausts and a variety of bodies with graceful lines. Nowadays the cars are extremely valuable.

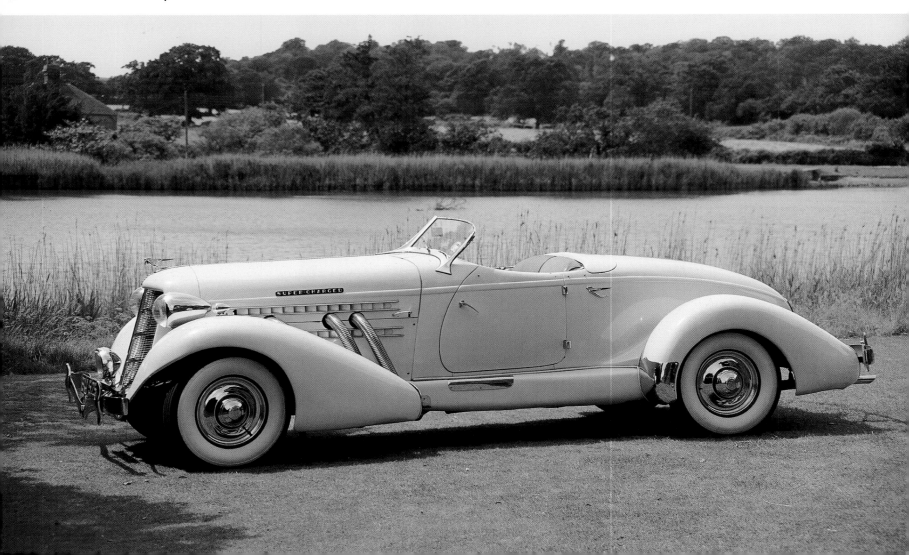

AUDI

This German marque, founded in 1909, became part of Auto Union in 1932, when they merged with Horch, DKW and Wanderer, but all four were located in what became East Germany and ceased trading after 1945. The Audi name was revived in 1965 when Volkswagen obtained the rights, and it soon expanded as a prestige marque. By the 1980s they were in the technical vanguard, especially with four-wheel-drive.

1980 AUDI QUATTRO This model was a major step forward for it introduced four-wheel-drive to a high-powered coupé designed for the road. It was to prove outstanding on ice or snow as well. In addition, its turbocharged, overhead-camshaft, 2.1-litre engine had five cylinders, rather than the usual four or six, a type the firm had first used in 1976.

1985 AUDI QUATTRO SPORT The new model soon added anti-lock braking to its specification, while later came a 2.3-litre engine, to be followed by further electronic controls over traction by torque sensing.

1992 AUDI CABRIOLET Saloons improved performance and reduced noise by using an aerodynamically designed shape. This work also went into the cabriolet to that model's benefit, although the folding top would detract from aerodynamic efficiency to a degree.

AUSTIN

Founded by Herbert Austin in 1906, this firm was based at Longbridge, near Birmingham, and became one of Britain's greatest producers. Initially they built conventional cars, but a one-model policy from 1919 was not a success. The famous Seven came in 1922 and turned the tide, after which they built sound models for volume sales. They merged with Morris in 1952 and are now part of today's Rover group.

1908 100HP AUSTIN In the Edwardian style, built for racing but fitted with road equipment, this car had a six-cylinder, 9.7-litre engine which lacked the power to keep the heavy machine up with its larger-capacity and lighter rivals.

1910 AUSTIN 7HP This model, actually a Swift with a different radiator, was typical of the early cars from the firm, being well made, reliable and conservative in its design. It had a four-cylinder engine but they also listed a single.

1931 AUSTIN SWALLOW The famous Austin Seven was announced in 1922 and built up to 1939 to offer a real car in miniature rather than a cyclecar. Many bodies were listed for the various versions and included this one by William Lyons' Swallow works. It gave owners his special line and style.

1935 AUSTIN 10 LITCHFIELD This model was typical of the larger saloons the firm built up to the mid-1930s in 10, 12, 16, 18 and 20 forms, all replaced by a new style around 1937. In this case the car is powered by a 1.1-litre, four-cylinder, side-valve engine of modest output.

1936 AUSTIN 7 RUBY This saloon became the most common of the Sevens, in place of the earlier soft-top tourer, but kept the 747cc side-valve engine which was hard pressed to haul the extra weight along. The gearing was lowered to cope but, overall, the model lost much of the sparkle of the original.

ABOVE: **1951 AUSTIN A90 ATLANTIC** Built and styled for the American market, this model was powered by a 2.7-litre engine and listed in saloon and convertible forms. The central front lamp was part of the package but the customers preferred six-cylinder or V-8 engines, although the four was a success in the Austin Healey.

BELOW: **1953 AUSTIN A40 SOMERSET** The first postwar A40 models were the Dorset and Devon with two and four-door saloon bodies respectively. Both had a 1.2-litre, four-cylinder, overhead-valve engine and a separate chassis and were replaced by this model in 1952, listed in four-door saloon and two-door convertible forms.

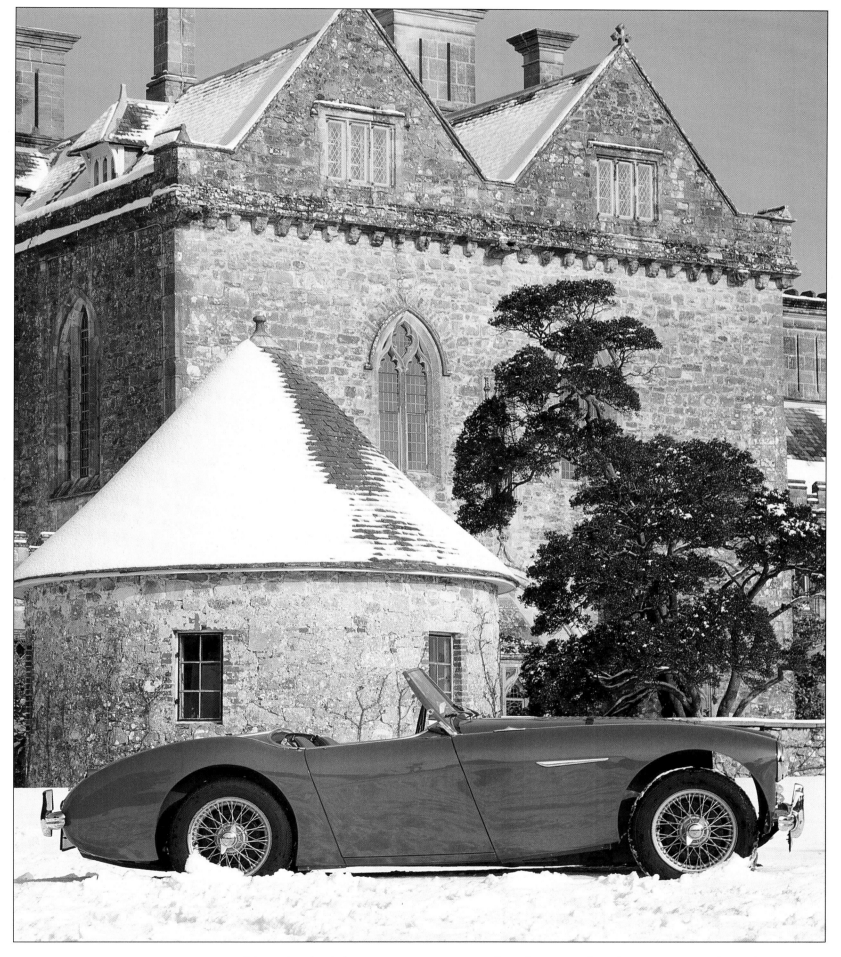

1956 AUSTIN HEALEY 100 Donald Healey built his own cars up to 1953 but then joined with Austin to produce these rugged, very popular sports cars for many years. Speed, dependability and a reasonable price ensured their success.

1958 AUSTIN HEALEY SPRITE This companion to the larger Austin Healey used a 948cc, overhead-valve engine able to run it up to 130 km/80 miles per hour. The one-piece bonnet-and-wing assembly lifted for access and the form soon led to the 'Frog-eye' name.

1959 AUSTIN A35 Following on the 803cc A30 introduced in 1951 came the A35 in 1956, fitted with the long-running 948cc A-series, overhead-valve engine. These were the first unitary construction (with body and chassis as one) Austins, and while they were successful in their decade, they were about to be eclipsed.

1959 AUSTIN A35 Typical flying-A bonnet motif used by the firm.

ABOVE: **1959 AUSTIN MINI SEVEN** Launched with a Morris version, the revolutionary Mini set new standards for carrying people and in road holding. A classless, ever-popular car, it changed small-car concepts for all time and remains in production.
BELOW: **1961 AUSTIN NASH METROPOLITAN** Built by Austin for Nash with American styling and by 1961 fitted with a 1.5-litre engine. It was produced for Nash from 1954 to 1962 and also with a Hudson badge for 1955-56.

ABOVE: **1965 AUSTIN
HEALEY 3000 MK III** Still
looking much as before, but with
more power, less noise and
improved interior, the MkIII
continued to offer 190 km/120
miles per hour performance at an
affordable price.

RIGHT: **1968 AUSTIN MINI
COOPER MK II** Raced and
rallied around the world with
enormous success, this version of
the Mini offered tremendous
performance on the road.

AUSTRO DAIMLER

At first an Austrian branch of the German Daimler firm, they became independent in 1906, recruited and enjoyed the services of Ferdinand Porsche, and then went on to build luxury and sports cars until taken over by Steyr in 1934.

1925 AUSTRO-DAIMLER
An elegant and advanced motorcar in the style associated with the firm for many years. Most built in the 1920s had an overhead-camshaft engine of six or eight cylinders, were well equipped, fast but expensive. The neat radiator cap and badge (left) confirmed the make and would be banned now.

AUTOCAR

Constructed in Ardmore, Pennsylvania, from 1901 to 1911, the Autocar was building cars having a twin-cylinder engine and shaft drive by 1902, thought to be on account of chain problems experienced in the New York-to-Buffalo Run of the previous year. For 1903 (the model below) the engine was rated as 10hp, while by 1905 most controls were sited under the steering column for ease of use.

BENTLEY

Throughout the 1920s the Bentleys were the archetypal British sports cars, large, powerful and five times winners at Le Mans. Falling on hard times, they were taken over by Rolls-Royce in 1931 and from then on become the more sporting version of that marque. In 1933 they introduced their 'Silent Sports Car', based on a Rolls although there were a number of significant variations, while postwar the two marques moved closer with few differences other than the radiator shell and mascot.

1926 BENTLEY 3-LITRE This example of the most popular of the early cars is a red-label, short chassis type fitted with a Vanden Plas body. Single overhead camshaft, four valves per cylinder and twin magnetos made for an expensive car. There were also blue and green label versions and the type won at Le Mans in 1924 and 1927.

1930 4½-LITRE BLOWER BENTLEY Brutally functional with its Roots supercharger mounted ahead of and driven from the crankshaft, this model was the epitome of the late-1920s cars. Only around 80 were built. Ettore Bugatti called them 'the fastest lorries in the world'.

1930 BENTLEY 8-LITRE The massive six-cylinder engine was typical of the marque and some fine saloon bodies were built on the chassis as well as this traditional sports type. A supercar of its day, large, robust, but not quite as heavy as it might seem.

1930 BENTLEY 8-LITRE Radiator cap and badge on a saloon, showing the flying-B to the world.

1935 BENTLEY 3½-LITRE This was the 'Silent Sports Car' which was similar to the Rolls 20/25 model but used a shorter chassis and had twin carburettors for the engine. With the firm under Rolls-Royce control, the Bentley became more a luxury sporting car rather than the outright sports machine of the past. Away went the overhead camshaft but in came more refinement.

ABOVE: **1948 BENTLEY MK VI** After the war, Bentley's 4½-litre, straight-six engine was fitted to a variety of bodies. Most used the Standard Steel saloon, but there were some custom built such as this lovely Mulliner drophead coupé. From 1951 the engine was enlarged to 4.6 litres.

BELOW: **1949 BENTLEY MK VI** This, another example of the first generation of postwar Bentleys, is fitted with a fine Sedanca de Ville body, rear wheel spats and a higher line for the headlamps. The chauffeur was expected to live with any passing rain showers.

RIGHT: 1949 BENTLEY MK VI ROYSTON A complete change and well removed from the usual saloons, coupés or limousines, this postwar open model was an unusual custom build. Its wire wheels and low line were more reminiscent of the prewar SS Jaguar than the big, open Bentleys of the 1920s.

BELOW: 1951 BENTLEY MK VI PARK WARD A very fine sporting two-door coupé body from one of the best British coachbuilders graces this car and, while it would have been custom built, it was a form that was not uncommon for the marque. Definitely a car for Derby Day, Ascot and Henley.

1955 BENTLEY S1 CONTINENTAL The Continental R of 1952 was a fastback replaced by this model which had saloon or drophead coupé bodies from a number of coachbuilding firms. The engine went up to 4.9 litres, and the marque moved closer to its 'sporting Rolls' role. Both Mulliner and Park Ward built two-door bodies.

1965 BENTLEY S3 CONTINENTAL While the S3 was perceived as a Rolls with a Bentley radiator, the Continental continued to offer a more sporting line with bodies from Park Ward, James Young or, as here, Mulliner. S3 models fitted a 6.2-litre, V-8 engine and all had the four headlights.

1988 BENTLEY TURBO R Through the 1970s Bentley was the poor relation to Rolls, despite the cars and their prices being virtually the same. The marque's own identity came back in the next decade thanks to models such as this which added a turbocharger to put the sports back into Bentley driving while retaining all the existing luxury and superlative finish.

BENTLEYS Old and new Bentleys in a contrast on the Members' banking at Brooklands. One is a 4½-litre blower model, an example of supercharging in the 1920s, the other a 1991 Turbo R with the modern equivalent. A fine demonstration of 60 years of progress in all departments.

BENZ (REPLICA)

It all began over a century ago when Karl Benz built a tricycle with a 954cc, single-cylinder engine and drove it on the streets of Mannheim in Germany. Later came four-wheelers with a 1045cc single-cylinder engine and Benz led the world in car production for a decade. This working replica (below) was built in 1989.

BMW

One of the great German marques originated as a manufacturer of aircraft engines and motorcycles. Bayerische Motoren Werke was formed from mergers in 1917. In 1923 they showed their flat-twin R32 motorcycle to the world and five years later took over an automobile factory. Since then, through both good years and bad, the quartered blue-and-white symbol has become a brand meaning technical excellence and high-quality production standards.

1937 BMW 328 SPORTS
Typical of the late-1930s series of models and one of the finest sports cars of its day, the 328 used a 2-litre, six-cylinder engine with an advanced overhead-valve layout, a tubular frame and had hydraulic brakes. A car capable of 160 km/100 miles per hour.

1958 BMW 507 In the 1950s the firm built a series of large, rather expensive cars powered mainly by a V-8 engine. Fine, elegant cars, available with a detachable hardtop. Production became limited, although the company was also making the Isetta bubble car from 1955.

1973 BMW 3.0 CSL The modern BMW began with the 1500 in 1962. The CSL evolved from it via the 2000 and 2800 as engine capacity increased over the years. It had a lighter body, wider wheels, a sports steering wheel and special seats and served as the competition model as the firm made its mark in production car racing.

1980 BMW M1 Developed for competition, the M1 had some of its roots in a 1972 experimental safety car. What finally evolved was a mid-engined layout that proved docile and comfortable to drive on the road, yet capable of some 255 km/160 miles per hour on the racetrack thanks to the 3.5-litre, six-cylinder, twin-cam engine. The chassis was by Lamborghini and the styling by Giorgetto Giugiaro.

1984 BMW 323i During the 1970s BMW introduced their series concept whereby one basic body size could be listed with a variety of engines, specifications and trims. Smallest was the 3-series which became the status company car of the 1980s with engines from 1.6 to 2.5 litres and four or six cylinders. The 323i has the 2.3-litre one with fuel injection.

RIGHT: **1986 BMW 635CSi** While the 5-series occupied the middle ground, the 6-series aimed higher and, with the 7-series, was intended to capture sales in the luxury and high-level executive markets. Top of the line was this 3.5-litre model which had the power, refinement and ability to cruise far and fast in comfort.

BELOW: **1987 BMW 325i** A 3-series convertible was created by adding bracing to the screen pillars while the hood was stowed out of sight under a panel. In this case the 2.5-litre engine was fitted which gave it a 215 km/135 miles per hour top speed, but 318i and 320i versions were also built.

ABOVE: **1990 BMW M3** Like the M1, the M3 was built for road and track. It had all the best features of the 3-series and a 2.3-litre engine able to push it to over 225 km/140 miles per hour, so went into volume production. It had all the required gear such as close ratios, limited-slip differential, well-controlled suspension and low-profile tyres while retaining most interior fitments.

LEFT: **1994 BMW ALPINA B3** Alpina was involved with BMW's competition efforts from the 1960s and eventually became a marque in their own right. The close connection led to the production of BMW-approved, uprated performance versions of the stock models.

BRUSH

Designed and built in Detroit by Alanson P. Brush, these cars (below, a 1909 Model B) used single-cylinder engines ranging from 6 to 10hp from 1907 to 1911, listing a twin for 1908 only. The cars were further out of the ordinary in using oak, hickory and maple woods for the axles and frame, while all four wheels had coil springs for their suspension. Simple enough to be reliable, the marque had a successful though brief life.

BUGATTI

Ettore Bugatti was an Italian artist and genius who built cars in France while supervising his many other activities and operations. Known as 'Le Patron', he controlled all and the cars he produced remain some of the most desirable of all time. From the Type 13 to the postwar Type 73, all were individually special.

1927 BUGATTI TYPE 37A
A Grand Prix car having a supercharged four-cylinder engine and the characteristic radiator. Normally fitted with wire wheels, this example has the famous alloy-spoke type.

1927 BUGATTI ROYALE TYPE 41 The ultimate in carriages, this creation was intended to be bigger and better than any other luxury car. Its straight-eight engine swept nearly 13 litres and the result weighed close to three tons. Only six were made, all for 'worthy' clients, and are now incredibly valuable. The radiator cap of the Royale (left), the elephant, was a gift that went with each car.

ABOVE: **1932 BUGATTI TYPE 55** A roadster capable of some 185 km/115 miles per hour, and which comprised the Type 54 chassis fitted with the Type 51 supercharged, twin-overhead camshaft, 2.3-litre engine. In the style of the master, elegance was allied to performance.

BELOW: **1933 BUGATTI TYPE 51A** Final version of the grand prix road-racing car with the supercharged straight-eight, twin camshaft, 1.5-litre engine and alloy-spoked wheels. Wonderful.

BUICK

Started in 1903 by a Scottish plumber, this marque became a founder member of General Motors in 1908. A major American firm, it was successful in the 1920s, dropped back and then revived again in the late-1930s. Postwar they continued as a major player, offering plenty of car for the money.

LEFT: **1936 BUICK 8** Typical of the body style adopted for the late-1930s, this massive limousine with its 5.2-litre, straight-eight, overhead-valve, engine was once used by Edward VIII.

BELOW: **1941 BUICK SUPER 8** A fine convertible body over the eight-cylinder engine was one of the many styles offered along with some custom builds. The 1942 cars were extensively restyled but the war delayed their wide distribution so they appeared fresh in 1945. Engines were of 4.1 or 5.2 litres and had overhead valves, producing from 115 to 165bhp.

ABOVE: **1953 BUICK SKYLARK** An early anniversary limited edition to celebrate 50 golden years. Only 1690 were built, and at $5000 they were twice the cost of the base cars. Revised to be less radical for 1954 with flared wheel arches and new rear end styling, sales dropped and so was the car.

BELOW: **1954 BUICK SKYLARK** Only 836 of these were built, powered by the 5.3-litre, V-8 engine producing 200bhp, but at twice the price of a base sedan and much less radical than earlier versions.

CADILLAC

From 1902 to today, a leading American firm that built many great cars in its time. They pioneered and demonstrated the concept of spares that had no need for hand fitting in Edwardian days, produced a V-16 engine in 1930, often led the way in style in the postwar years and, as part of General Motors, became THE American luxury car.

LEFT: **1929 CADILLAC DUAL COWL** Typical of the cars built in the late-1920s, powered by a large V-8 engine, and well able to cope with America's distances while retaining a sophisticated big-city style.

BELOW: **1931 CADILLAC 370A** As the Depression struck, Cadillac brought out not one but two of its most prestigious models. This was actually the smaller which had to manage on just 6-litres from its V-12 engine, under the bonnet of the elegant Fleetwood body.

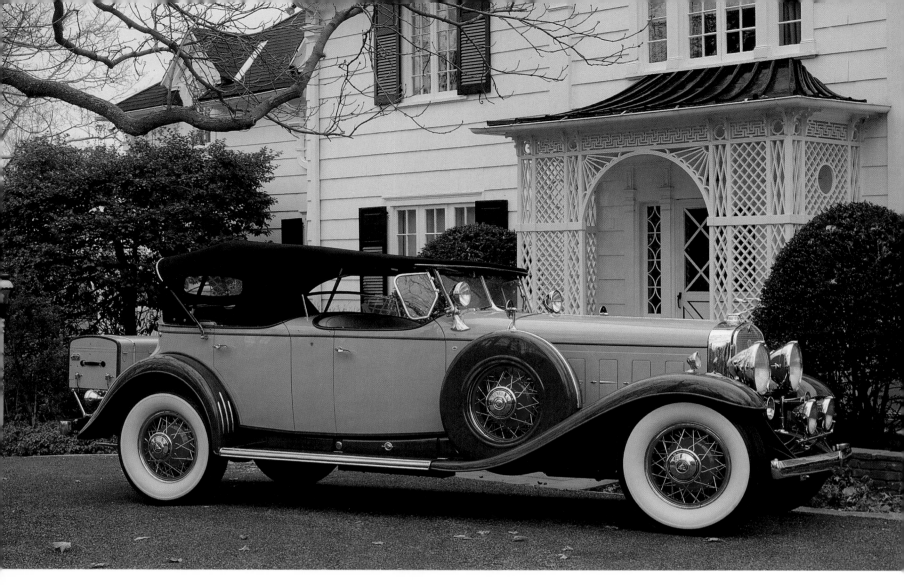

ABOVE: **1932 CADILLAC 452B** The fabled V-16 whose 7.4-litre engine had its two banks of eight cylinders at a narrow 45-degree angle, plus overhead valves to push the big car along at close to 160 km/100 miles per hour. Available with any body, stock or custom, all with great presence to maintain the firm's eminence through the hard times.

BELOW: **1947 CADILLAC** Early postwar cars continued the prewar themes and all used the same 5.7-litre, V-8 engine from those times. The next year the tailfin era began as styling was derived from the Lockheed P-38 Lightning aircraft.

ABOVE: **1950 CADILLAC** For 1949 Cadillac switched to a 5.4-litre, overhead-valve, V-8 engine, better and lighter than the old side-valve unit, and the tailfins appeared. The next year brought the dummy air intakes.

LEFT: **1955 CADILLAC FLEETWOOD** By the mid-1950s the tailfin theme and other features had evolved to this point, while the power output from the V-8, overhead-valve engine rose for most years. The front bumper (below) had more pronounced features as time passed.

OPPOSITE: **1958 CADILLAC ELDORADO BIARRITZ** The Eldorado first appeared in 1953, a two-seater convertible whose soft top could be stowed under a deck panel out of sight. Expensive at $7500 it was joined by the Brougham Sedan which ran to $13,000.

ABOVE: **1958 CADILLAC ELDORADO BIARRITZ** The frontal aspect of the Biarritz and its four headlamps.

LEFT: **1958 CADILLAC ELDORADO BIARRITZ** Tailfin and rear quarter detail, the fins having grown over the years.

ABOVE: **1959 CADILLAC COUPÉ DE VILLE** The tailfins reached their most extreme in 1959, after which they shrank although the rear deck remained a vast area for some time. The engine, used for many years, remained the 6.4-litre, V-8 producing 325bhp.

CATERHAM

Based on the early Lotus Seven, which became famous thanks to the cult TV series, *The Prisoner*, the manufacturer was taken over by Caterham Cars and continues to this day (below, a 1994 Super 7). It offers the same stark emphasis on performance as the original Colin Chapman design.

CHEVROLET

Founded in 1911, this American firm became the mass-market producer for General Motors, in competition with and usually outselling Ford. The high level of production gave buyers plenty of car for their money.

1953 CHEVROLET CORVETTE A sporty supplement to Chevrolet's traditional emphasis on conventional sedans, the Corvette had a two-seater, glass-fibre body. From this first model, year on year, it developed and introduced new styles and more power to keep it in the vanguard of American performance cars.

1957 CHEVROLET BEL AIR
First seen in 1950 as a hardtop coupé fitted with a six-cylinder engine, the Bel Air soon became a top-of-the-line range. Powered by a 4.6-litre, V-8 engine that produced 185bhp, with options ranging up to 283bhp, it was listed in seven forms in 1957 and ran on to 1971.

1958 CHEVROLET IMPALA
Introduced as part of the Bel Air range in 1958, the Impala became a range of its own the next year, offering greater comfort and running up to 1971.

1959 CHEVROLET CORVETTE Six years of development brought the Corvette to this point with dual headlights, a bigger V-8 engine and more power. The result was a new production record promptly beaten the following year.

1969 CHEVROLET CAMARO Created as a sporting compact, the Camaro had an extensive range of options listed for it to provide any combination of performance, luxury and economy to suit the buyer. It proved a big sales hit.

1969 CHEVROLET STINGRAY First listed in 1963 as the Sting Ray, this represented a near-total revision for the Corvette. Always dramatically styled, it ran to 1977, and offered considerable performance. This model had the 7-litre, V-8 engine pushing out 390bhp to 435bhp, depending on tune.

ABOVE: **1981 CHEVROLET CAMARO Z28** Given a facelift for 1978, the V-8 powered sports coupé was at its most expensive in Z28 format but had engine options including four-barrel carburettor or fuel injection.

BELOW: **1987 CHEVROLET CAMARO IROC-Z** The International Race of Champions, or IROC, was a race series for top drivers all using Camaros. This model celebrated that with a performance package to uprate engine power, suspension and road holding.

CHRYSLER

Founded in 1924, this firm may be the smallest of the American 'Big Three' but is still a high-volume producer. Designs were at first advanced and prestigious, then radical in the 1930s, before a lengthy conservative period. Radical returned in the mid-1950s. The 1970s were troubled times for Chrysler, although they bounced back in the next decade to continue to nag at Ford and GM.

LEFT: **1930 CHRYSLER IMPERIAL** A classic car, powered by 5.1-litre, straight six, side-valve engine mounted in a long-wheelbase chassis. Many body styles were offered, both from the factory and from the best of the American specialist coachbuilders.

BELOW: **1932 CHRYSLER IMPERIAL** One of America's classic cars with a magnificent appearance in any of its many body styles. A straight-eight, 6.3-litre engine pulled some two tons of car along at a good speed but with a considerable thirst.

1941 CHRYSLER NEW YORKER Priced lower and on a shorter wheelbase than the Imperial, the New Yorker models were listed in a variety of body styles and used the straight-eight, 5.3-litre engine.

1940 CHRYSLER NEWPORT In 1940 the firm designed two 'Ideal Cars', the Thunderbolt hardtop and this Dual-Cowl Phaeton, and built six of each. The Newport used the Crown Imperial chassis and 5.3-litre, eight-cylinder engine producing 143bhp. One of these was the pace car for the 1940 Indianapolis 500 while the concealed headlamps and front style were similar to Cord and Lincoln.

1948 CHRYSLER TOWN & COUNTRY This curious estate car, known as a 'woody' in America, was built as a sedan or convertible. Listed from 1941 to 1950, one version carried up to nine people.

1955 CHRYSLER C-300
This model allied performance to style by using the powerful hemi-head, V-8 engine and a line evolved from Ghia-bodied show cars. With 300hp from the start, the car was quick and broke with the firm's conservative image.

1956 CHRYSLER NEW YORKER The hemi-head V-8 engine went into other models as well as the 300 series, in this case stretched out to 5.8-litres and offering a modest 280bhp, 75 down on the best option.

1957 CHRYSLER 300C Listed as a hardtop or convertible coupé, the 300-series cars were up to 6.4-litre and 375bhp as standard, 390 with the performance pack. The body gained modest fins and its own grill form while the TorqueFlite automatic transmission was controlled by push buttons on the dash.

CITROËN

This major French company was founded in 1919 and quickly became a leading producer of low-priced, reliable and conservative cars. This was enhanced in the 1930s when their natural flair moved them into the vanguard of design and style, this continuing postwar with both highly advanced and very basic, but idiosyncratic, designs, all with their unique fashion. Later they became part of Peugeot and less radical.

1922 CITROËN B2 Introduced at the same time as the Type C 'Cloverleaf' of 855cc, the B2 had a 1453cc side-valve engine and three-speed gearbox to give it correspondingly more performance. They were imported to Britain as parts and assembled at Slough, the steel saloon bodies jig-built from pressings.

1938 CITROËN 11CV The leap forward to *traction avant* came with the 7CV in 1934, combining front-wheel-drive and unitary construction with the mechanics of engine and gearbox bolted to the front of the body shell. The 11CV's overhead-valve engine was stretched to 1911cc. The car had incredible road holding.

1959 CITROËN 2CV First seen in 1948 and destined to become a cult car, the 2CV had a flat-twin, overhead-valve engine, ranging from 375 to 602cc. There was interconnected suspension allied to some awesome roll angles on corners to provide minimalist motoring in a car that was listed right up to 1990.

1966 CITROËN DS21

This French combination of revolutionary, self-levelling hydropneumatic suspension, power gearshift, clutch and brakes (the latter controlled by a tiny pedal with little movement) with their prewar 1911cc engine caused a sensation when first seen in 1955. Built in saloon, drophead coupé and estate forms with engines of up to 2347cc until 1975.

1975 CITROËN SM

A 1970 combination of Citroën suspension and other technology with a Maserati four-camshaft, V-6 engine and five-speed gearbox under the established French body shape led to this version with fuel injection that could reach over 210 km/130 miles per hour.

1988 CITROËN BX 19 GTi

There was more convention under Peugeot control although the 2CV hung on to 1990 and the GSA hatchback was true to Citroën form. However, the AX, BX and CX series adopted group engines and other benefits while retaining French chic style and some Citroën quirks.

CORD

An American marque of the 1930s which opened for business with its advanced and expensive L-29 just as the stock market crashed. It had a 4884cc straight-eight, side-valve engine, front-wheel drive, hydraulic front brakes, the front ones inboard. The firm closed down but returned in 1936 with an equally unconventional design which was only built for that year and the next.

1937 CORD 812 The L29 seemed conventional, but had front-wheel-drive which allowed for a lower bonnet line. The 1936 car retained the drive system but added highly radical styling with alligator bonnet, slatted grill, manually retracted headlamps and an aircraft-style dash. Under the hood went a 4730cc, side-valve V-8 engine which drove a four-speed overdrive gearbox with electro-vacuum change. The result was the 810. By adding a supercharger, and the plated exhausts, the 812 was created, well able to exceed 100mph. Both were built with sedan (right), berlin, convertible coupé (centre) and phaeton (bottom) bodies.

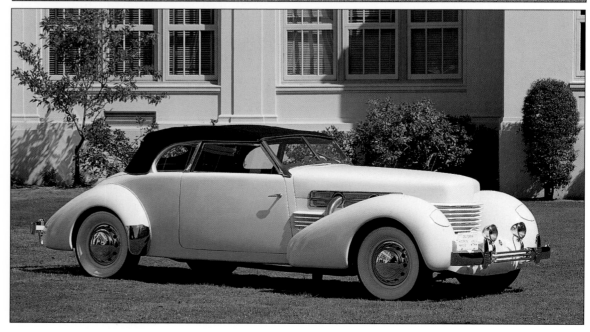

DAIHATSU

This long-established Japanese firm did not build a car until 1958. They specialised in the small and off-road categories so were well placed to field this leisure vehicle (below, the 1994 Daihatsu Sportrak) with its 1.6-litre engine and four-wheel drive which made it capable on tarmac, on tracks and in the field.

DAIMLER

Car production began at this firm in 1897 and has continued since, although the make went through various hands over the years. Much used by the Royal Family between the wars, when it produced luxury and high class machines, it faded postwar and was taken over by Jaguar in 1960, since when it has been an upmarket badge for that marque.

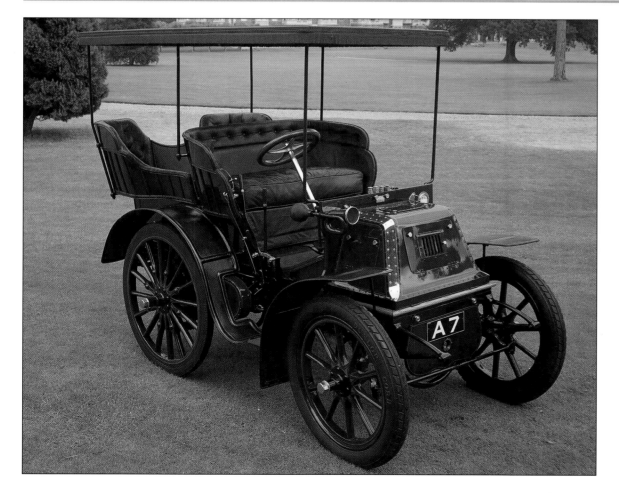

1900 DAIMLER Edward VII, when Prince of Wales, was an early passenger in a Daimler, and bought a 6hp model in 1900 – the first of the Royal customers.

ABOVE: **1939 DAIMLER DB18** A sports tourer produced alongside the massive limousines, powered by a 2522cc straight-six, overhead-valve engine and fitted with a fine dual-cowl phaeton body. A good performer built in small numbers in the British style of the period.

BELOW: **1963 DAIMLER V250** Edward Turner of Triumph motorcycle fame designed a fine V-8 engine thanks to the link to Daimler via BSA. Here it was mated to the MkII Jaguar body to produce something a little special for the discerning customer.

DARRACQ

This 1904 Darracq (below) is 'Genevieve', the star of the film that brought the veteran movement to the general public's notice. The French firm was already eight years old when this 8hp model was made. In 1920 it merged with Sunbeam-Talbot, from which a confusion of marque names emerged.

DATSUN

This Japanese firm dates back to 1912. An early association with Austin was resumed after 1945, but then came a line of conservative saloons. The 1970s brought the successful 240Z sports coupés which became the 280ZX (below, the 1982 280ZX) in 1978. In the early 1980s the company began marketing its cars as Nissans.

DE DIETRICH

Also listed as the Lorraine Dietrich, De Dietrich built locomotives before adding cars in 1896 (below, a 1903 De Dietrich). After the end of the First World War, their home in Alsace was returned to French control by Germany. De Dietrich continued to make cars until 1935, winning Le Mans twice in the 1920s.

DE DION BOUTON

The combination of Marquis de Dion and Georges Bouton put this French firm at the head of the industry in its earliest days when it built vehicles itself and supplied thousands of engines to other manufacturers. A single-cylinder engine, simple two-speed gear and ease of control made the cars most successful, and the firm possibly the largest car producer of its time (below, a 1901 De Dion Bouton Vis-à-Vis).

DELAGE

From a modest 1905 start, Louis Delage moved to luxury and successful racing cars between the wars. French elegance continued after Delahaye took them over in 1935 and fitted the Delage engine in their own bodies but both names vanished from the market in the 1950s.

1924 DELAGE BOAT-TAIL
The 1920s were a significant time for the firm and one when many coachbuilders applied their skills to the Delage chassis. This example was listed as the CO-series, which fitted a six-cylinder engine, and has a body by Labourdette featuring dual windscreens. Detail (below) of the bound steering wheel, instruments and controls before the driver.

1937 DELAGE D6-75 Typical of the cars of the late 1930s where the Delage 2800cc, six-cylinder, overhead-valve engine was fitted in a Delahaye drophead coupé body. It combined the good running of the engine with the chic and elegant French style.

DELAHAYE

A French firm that dated from 1894 but whose best years came in the late 1930s. Early models were mainly dull tourers but then came some real French style and elegance which continued postwar to 1954 when the cars were dropped due to the penal French taxes on luxury vehicles.

RIGHT: **1938 DELAHAYE T135M** Superlative body from Figoni et Falaschi, one of the French coachbuilders who supplied Delahaye customers. The car would be seen prewar, parked in the Champs Elysées or on the Côte d'Azur. Underneath went a 3557cc six-cylinder, over-head-valve engine, good brakes and decent handling which reflected the firm's grand prix involvement. There was also the T165 which fitted a 4480cc, V-12 engine developed from the grand prix cars.

BELOW: **1947 DELAHAYE 135M** The postwar cars still used the prewar 3557cc six-cylinder, engine and had the Cotal electromagnetic gearbox offering four speeds plus reverse and controlled from a miniature gate on the steering column. The bodies varied but this Vanden Plus drophead coupé is typical.

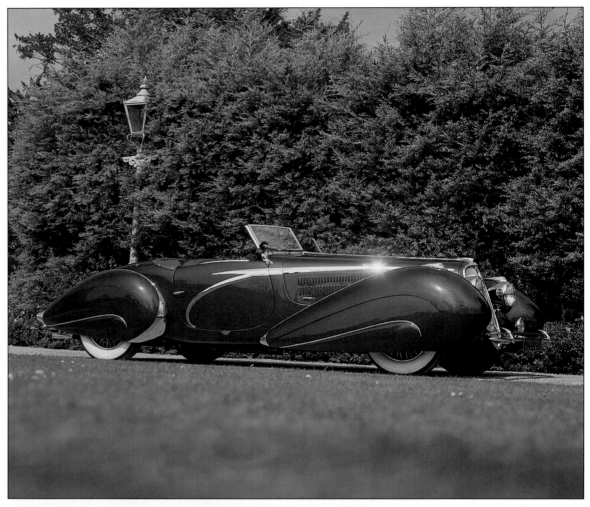

DELAUNEY-BELLEVILLE

An early French firm who built conservative chassis for their wealthy buyers who ordered the very finest of coachwork in the days before the First World War (below, a 1909 model). From 1908 the Tsar was their best customer and headed a long list of notables. Their prestige fell away in the 1920s but they survived to 1950.

DELOREAN

Setting up a car plant in Northern Ireland, John DeLorean had Lotus design the chassis and Giugiaro the styling, the result powered by a Renault V-6 engine (below, a 1982 DMC-12). Production quality and other problems sealed the car's fate, but it found fame as a time machine in the *Back to the Future* films.

DESOTO

Launched in 1928 as a cheaper Chrysler, the DeSoto line was conventional, even dull at times, but improved during the 1950s (below, a 1958 model). The big fins, a good V-8 engine and the extensive front trim helped for a while, but by 1961 the name had been squeezed out by other group products from Dodge and Chrysler.

DE TOMASO

Alejandro de Tomaso came from Argentina, raced cars and became a dominant figure in the Italian car and motorcycle industries. The Panterra appeared in 1970 using a Ford V-8 engine under a Ghia-designed body (below, a 1975 Panterra GTS); but lack of reliability, body corrosion and a high price kept sales low. However, it thundered on into the 1990s with a 275 km/170 miles per hour potential.

DODGE

Founded in 1914 and bought by Chrysler in 1928, this American firm has always offered value for money. During the 1930s, Dodges had to compete against other cars from Chrysler, but their range broadened after the war to follow industry trends. Along with the usual sedans and convertibles there were sports coupés and other high power models making Dodge the performance marque in the Chrysler corporation.

LEFT: **1934 DODGE CONVERTIBLE** These models were similar to the Plymouths of the time and had independent front suspension. They were also built in sedan form and all had a choice of various six-cylinder, side-valve engines.

BELOW: **1959 DODGE CUSTOM ROYAL** This is the Lancer hardtop coupé version of a popular series powered by a V-8 engine, here sporting fashionable fins. Priced at the top end of the Dodge range, these cars sold well from 1955 to this final version.

1969 DODGE CHARGER DAYTONA This series was first seen in 1966 as a hardtop coupé in the muscle-car league and this version was added for 1969 to compete in American long-distance races. Only 505 were built, just enough to qualify as a production car, and the result was a success. The distinctive, high-level wing on the Daytona was matched by a bullet nose and hidden headlights at the front.

1970 DODGE CORONET SUPER BEE Listed from 1968, this model fitted a 335bhp, V-8 engine as stock with a 390bhp option for 1969 and a 425bhp option for this, its final year. It retained the high wing but had a more conventional and heavy front end trim.

1993 DODGE VIPER Able to compete with the fastest sports cars in the world, the Viper used an 8-litre, 90-degree, V-10 truck engine to power it to 270 km/167 miles per hour in the fifth gear of its six-speed box. Front engine, rear-wheel drive, and two seats all added up to the classic format with road-holding, handling and brakes to match.

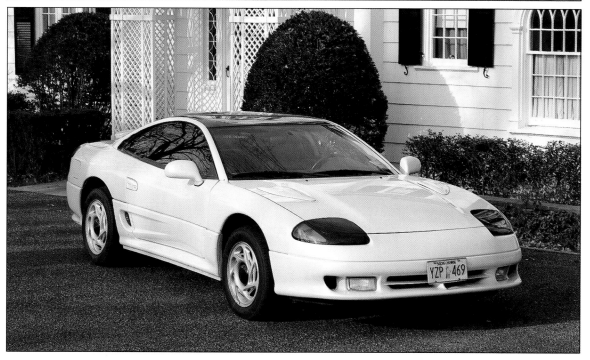

1994 DODGE STEALTH R/T TURBO For buyers who preferred a coupé, Dodge offered the Stealth in several forms. All used a 3-litre, twin-cam, 24-valve, V-6 engine and front-wheel drive. This one had four-wheel drive, four-wheel steering and an engine with twin turbochargers, while all models had all the modern amenities. The resulting package offered 255 km/160 miles per hour and close to 1g cornering potential.

DUESENBERG

Fred and August Duesenberg founded this firm in 1920 and produced the Straight-8 model from 1921 to 1927 with few changes, but early sales were poor despite the advanced specification, high quality and excellent workmanship. Erret Lobban Cord took over and they then developed one of the great classic luxury cars of all time, the Model J which was built up to 1937.

THIS PAGE AND OPPOSITE: **1930, 1931 & 1933 DUESENBERG MODEL J** One of the finest cars of its day which compared with the best in the world, having sophisticated engineering, the smooth running of the luxury car but the sporting appeal that ensured that most, but not all, had an open or convertible body. Massive in concept but invariably elegant in its size, the cars were powered by a 6882cc straight-eight engine having twin overhead camshafts and four valves per cylinder built by Lycoming, another Cord company. The massive frame had six tubular cross-members, hydraulic shock absorbers, hydraulic brakes with servo assistance while the transmission was via a twin-plate clutch and three-speed gearbox to a rear axle available in four ratio options. Bodies came from the leading firms and the result was a very expensive car.

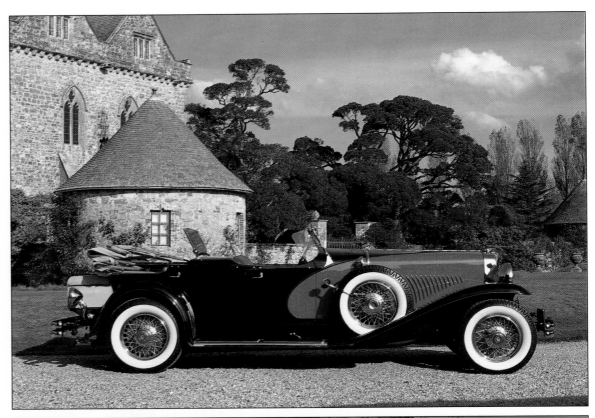

1933 DUESENBERG SJ The fabulous J became more so in 1932 when the SJ version with centrifugal supercharger arrived. The engine was strengthened to cope with its 320bhp, close to 210 km/130 miles per hour was claimed, the front springs were stronger and the price went up. While over 400 Model Js were built, there were just 36 SJs, most of which went to celebrities.

EDSEL

Ford's folly, or a good car at the wrong time – two views of the Edsel which came in 1958 (below, a 1958 Citation) and went two years later. Named after the only son of founder Henry Ford, the car had its own style, especially the horse-collar grill. Sadly, it arrived at the wrong time and failed so is now collectible.

EXCALIBUR

This firm was set up in 1964 to build classic cars styled on the 1928 Mercedes-Benz SSK model and became a success story from the start. Carefully assembled, exclusive and expensive, production numbers were always minimal, but the buyers had cars which were fast, powerful and totally different.

1965 EXCALIBUR SSK With 290bhp from its V-8 Studebaker engine, sitting in a convertible chassis from the same firm, there was an amount of re-engineering needed to ensure safe handling on the road, and the result was a hit. The flexible exhaust tubing came from the German firm that had originally supplied Mercedes.

1976 EXCALIBUR SS The first series became the series II in 1970 and III in 1975, moving on to Corvette engines from General Motors and then to a big Chevrolet V-8. The chassis also changed and both manual and automatic transmissions were listed along with roadster and phaeton bodies.

ABOVE: **1976 EXCALIBUR SS** The phaeton body in the Series III.
BELOW: **1981 EXCALIBUR SERIES IV ROADSTER** The next series opted for another General Motors V-8 engine and automatic transmission. This roadster gained a rumble seat in the late-1930 Mercedes style.

EXCELSIOR

A Belgian firm in business from 1901 to 1932 built this classic tourer (below) in 1911 using engines with four or six cylinders There were overhead valves for the sporting versions in place of the usual side valves specified for the tourers. In the 1920s they built another fine series but in time their designs became dated. This cannot detract from the style and lines of the early cars.

FACEL VEGA

After supplying car bodies to other firms, Facel Vega became the last builder of luxury French cars with a Grand Tourer launched in 1954 using a Chrysler V-8 engine. Later came this coupé (below, 1960 HK500) with a choice of 5.9-litre or, as here, 6.3-litre engines and disc brakes all round from 1960. An attempt to produce their own, smaller, engine was a major factor in the firm's 1964 failure.

FERRARI

Exciting, exotic, expensive, extravagant all apply to the cars from Maranello that carry the prancing horse logo. Enzo Ferrari raced cars, ran a race team and began building road cars carrying his name in 1947. All were sophisticated, all were stylish, most were red and this tradition continues to this day. While Ferrari produce the chassis, the bodies came, and come from the leading Italian coachwork firms to complete a fine combination. All speak of speed, luxury, road holding and Alpine passes in every line. The takeover by Fiat in 1969 made no difference.

1949 FERRARI 166 Under the bonnet lies the Colombo-designed V-12, overhead camshaft engine of 1995cc driving a five-speed gearbox to make it go. Big drum brakes stopped it, the front suspension was independent and the fine coupé body was by Carrozzeria Touring.

1951 FERRARI 195 INTER This model was only built for one year and fitted the larger 2341cc version of the V-12 engine. In this case the body was a two-tone coupé but not in the usual red and by Vignale, one of the leading firms who supplied for this model.

ABOVE: **1956 FERRARI 410 SUPERFAST** Special 410 fitted with a Pininfarina body and shown at the 1956 Paris show. There were 24 sparking plugs for the 4.9-litre, Lampredi-designed, V-12 engine, the car had a shorter chassis than usual and the body tail fins were not a normal Ferrari feature.

BELOW: **1957 FERRARI 410 SUPERAMERICA** There were three series of this model built in small numbers over four years. All used the 4.9-litre engine and four-speed gearbox in a car that was large and heavy by Ferrari standards. Mainly sold to the rich and famous in the United States, they suited the long American roads.

LEFT: 1962 FERRARI GTO
The 250 GTO was built for sports car racing in the early 1960s and is perhaps the most desirable of all Ferraris. It won at Le Mans, had a Scaglietti body and was both very fast and very tough. This special version has a 4-litre engine.

BELOW: 1964 FERRARI 250 LM
The 250 GT series ran for a decade from 1954 but the LM, or Le Mans, was a new mid-engined concept built for racing, but street legal. It was powered by a 3.3-litre, V-12 engine and had fuel tanks on each side, just behind the doors.

ABOVE: **1973 FERRARI DAYTONA 365 GTB** First of the road cars to have twin overhead camshafts for a V-12 engine, the Daytona was Ferrari's fastest and most expensive production car up to that time. It changed little in its four years, this the last.

BELOW: **1973 FERRARI DINO 246 GT** The Dino was to be the less-expensive model, involved Fiat and did not carry the Ferrari badge. However, there was no mistaking the Maranello origins of the 2.4-litre, V-6 engine which was installed transversely aft of the seats. The result was a joy to drive.

ABOVE: **1973 FERRARI DINO 246 GTS** For this Dino model the roof centre section could be removed for open-air motoring while retaining the fine balance of the series. In the 1970s the engine grew to be a 3-litre V-8 but all had twin overhead camshafts.

ABOVE: **1978 FERRARI 512 BB** In 1973 Ferrari introduced their first car fitted with a flat-12 or boxer engine. It had twin overhead camshafts and was located just ahead of the rear axle under a body built by Scaglietti. The car became the 512 BB in 1976 when the engine was enlarged.

BELOW: **1985 FERRARI TESTAROSSA** In English 'Red head', this model name revived memories of a competition car of the late 1950s but in this case meant a flat-12, twin-cam, four-valve engine and highly-dramatic styling.

ABOVE: **1985 FERRARI 288 GTO** A very special Ferrari which revived a revered code and sold on this alone. It was fully equipped, a road-legal race car, able to exceed 300 km/185 miles per hour.

LEFT: **1988 FERRARI F40** A celebration of 40 years of production listed for six , the F40 had a twin-cam, V-8 engine and was an old-style, road-legal racer. High technology, noisy, but good for 320 km/200 miles per hour with that massive rear spoiler.

FIAT

One of the great Italian firms, launched in 1899 as Fabbrica Italiana Automobili Torino, who concentrated on building cars for the budget end of the market. Some models became classics. The firm remains a giant in Italian industry, now in control of several other Italian car firms.

RIGHT: **1913 FIAT TIPO ZERO** This was the firm's first mass-produced model, powered by a 1846cc, four-cylinder engine, and it set them on their path of building sound, reliable cars at an affordable price.

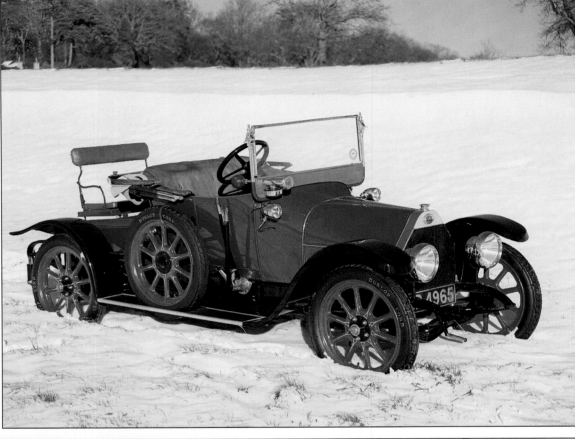

BELOW: **1935 FIAT 508S BALILLA** The 508 was first built in 1932 with a 995cc side-valve engine but grew into this overhead-valve sports model having four speeds, wire wheels and this stylish body with a small finned tail.

ABOVE: **1937 FIAT 500 TOPOLINO** A legendary Fiat, known as 'Mickey Mouse', and offering two seats, a 569cc four-cylinder, side-valve engine ahead of its radiator, four speeds and independent front suspension for minimal motoring.

BELOW: **1970 FIAT DINO** Before Fiat took over Ferrari the two worked in conjunction, each to introduce a Dino in 1966, both using the Ferrari engine and gearbox. However, the Fiat had the engine at the front and rear-wheel drive in conventional style, and was built up to 1972, fitting the 2419cc engine from 1969.

FORD

Henry Ford put the world on wheels when he introduced the immortal Model T late in 1908. It transformed motoring from an expensive pastime for the wealthy to a siginificant mode of transport for the masses. The company has continued along that line in America, Europe and Australia to this day.

RIGHT: **1913 FORD MODEL T** The combination of a 2.9-litre unstressed, four-cylinder, side-valve engine driving an epicyclic two-speed and reverse gearbox in a high-built and strong chassis made for a cheap, reliable car that was easy to drive and maintain and well able to go most places on the roads of its time. Always with rear brakes only, the 'Tin Lizzie' was built from 1908 to 1927 in the United States and also assembled in Manchester, some 15 million being produced in all.

BELOW: **1930 FORD MODEL A** The Model A replaced the T for 1928, having a conventional three-speed gearbox but remaining tough, cheap and reliable. Only built for four years but, unlike the T, it had brakes on all four wheels and kept the ability to operate on poor roads or cross-country.

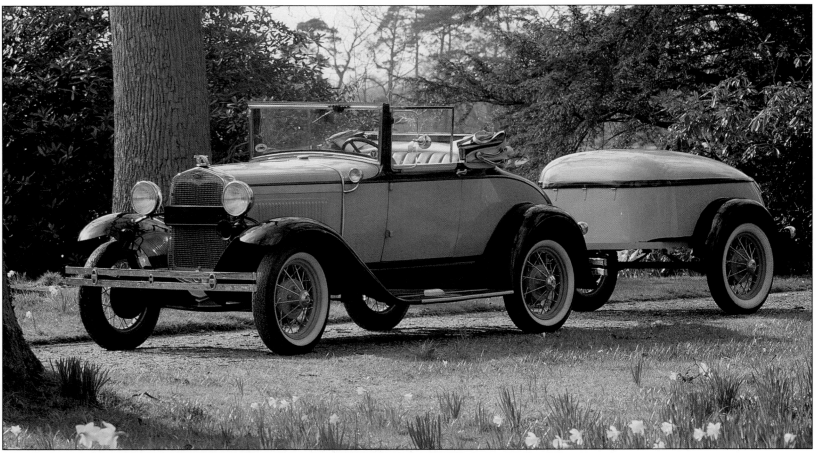

1936 FORD V-8 WOODY
The V-8 series was another Ford success from 1932 right through to 1953 in its side-valve form. Of 3.6-litre, it went into many body styles where its power made for a fast car and its strength and toughness ensured reliability.

1950 FORD CRESTLINER
This was a limited-edition model built for 1950-51 and based on the Tudor sedan but using the side-valve, V-8 engine rather than the usual straight six. The Crestliner was distinguished by its two-tone finish and a padded vinyl roof, while the 1951 cars had a new grill with small twin bullets on a heavier crossbar.

1955 FORD CONSUL MK I
Unitary construction, overhead valves, MacPherson-strut front suspension and hydraulic brakes moved Ford into modern times in 1950. The model ran on to 1962 in MkII form. This MkI convertible performed adequately on its 1508cc.

1956 FORD THUNDERBIRD
Introduced by Ford in 1955 to sell against the Chevrolet Corvette, the Thunderbird had a similar long model run. The early Ford was the better-built car and greatly outsold the General Motors product so was little altered for this model, except for an optional, larger V-8 engine.

1959 FORD GALAXIE SKYLINER The Galaxie series was introduced during 1959 to be Ford's top range and took over the convertible which was listed as the Fairline 500 Skyliner, their first mass-produced retractable hardtop. It came with the V-8 engine as standard.

1959 FORD GALAXIE SKYLINER The hardtop in mid-retract, on its way to be hidden under the rear deck.

1961 FORD ZEPHYR II The Zephyr began as a Consul with two more cylinders, and was both popular and able to run fast when tuned. The MkII came along in 1956 using a 2553cc engine and the name survived to 1972 in MkIV form along with the Zodiac version which had better fitments and trim.

1969 FORD MUSTANG SHELBY GT350 The Mustang was a major success for Ford partly because of the many options which enabled the buyer to range from pure economy to drag racer via luxury. Carroll Shelby took it a major step on to production racing in 1965, his version selling over 2000 each year from 1966 to 1969.

1966 FORD GT40 This GT40 is a street-legal version of the racing model Ford used at Le Mans. A 4.7-litre, V-8 engine pushed it to over 240 km/150 miles per hour.

1978 FORD MUSTANG II KING COBRA This was an attempt to ape the successful Shelby Mustang by adding decals and stripes to a car that performed adequately rather than as a class leader. The many options remained and this version proved to be the collector's choice.

1986 FORD ESCORT RS TURBO The old Escort was given front-wheel drive in 1980 and the new car was soon a favourite. The XR3 was the popular sports version but quicker still after 1984 was this one which added a turbocharger and other useful fitments for speed and handling.

RIGHT: **1986 FORD CAPRI 1.6 LASER** A car of the 1970s that refused to fade away despite dated looks. Devotees still loved its style and manners although hothatches outperformed it.

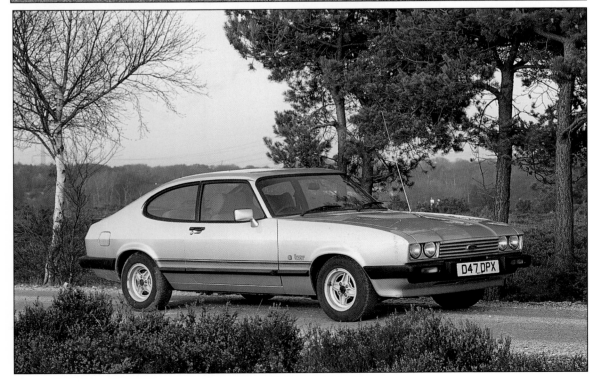

ABOVE: **1987 FORD SIERRA RS500 COSWORTH** The standard three-door Sierra was a poor seller at first, but powered up as the XR4i, and then with the Cosworth engine, it flew. The massive rear spoiler proclaimed its speed and in this form the specification was aimed at competition. Stock models ran to 240 km/149 miles per hour.

LEFT: **1988 FORD RS200** A road-going rally car using the turbocharged Cosworth 1.8-litre, mid-mounted engine and four-wheel-drive. Very quick as stock and much quicker in full competition form, but listed fully trimmed for road use for the small number made.

FRAZER NASH

Before the war this English firm built a range of stark sports cars whose solid rear axle was chain driven from a countershaft. Each gear ratio, including reverse, had its own chain plus sprockets, and was engaged by interlocked dog clutches. The lack of differential called for a special driving technique and the chain whir and outflung oil were certainly acquired tastes, but the cars worked wonderfully well. Postwar versions acquired a greater refinement.

RIGHT: **1933 FRAZER NASH TT REPLICA** Over eighty of these fine cars were built, a big number for the firm, using both four-cylinder and six-cylinder engines having overhead valves or overhead camshaft. They had four speeds, a quick clutch and a need to be powered round corners to force the rear end into place. Then they went where pointed.

BELOW: **1952 FRAZER NASH TARGA FLORIO** Postwar, the firm used the 2-litre, six-cylinder Bristol engine. Production remained minimal but the transmission became conventional while this model had a new, tubular frame and light-alloy body. Good for 175 km/110 miles per hour with basic engine, more from the tuned version.

GRAHAM

The three Graham brothers moved into cars in 1927, built close to 80,000 in 1929 and almost disappeared in the Depression. While their late-1930s cars had radical styling, the early were conventional. This model, a 1931 Custom Eight (below) had a 4.9-litre, side-valve, V-8 engine driving a four-speed transmission, although the 1930 cars were of 5.3-litre. Production ceased in 1941.

HEALEY

Donald Healey built cars carrying his name up to 1954 along with his involvement with Austin and Nash. He used the 2.4-litre Riley engine for a coupé and the open Silverstone models as well as this elegant 1947 Elliot saloon (below) which could run to 175 km/110 miles per hour. A pedigree car.

HILLMAN

A cycle firm that turned to cars in 1907 and built many modest but good models over the years. In 1963 it introduced the Imp (below), which had an 875cc, overhead-camshaft, all-alloy, four-cylinder engine at the rear giving good performance. The rear weight could affect the handling, and there were too many minor problems, but it ran to the end of the name in 1976.

HISPANO-SUIZA

This firm was established in 1904 but its roots in Barcelona went back to 1898. In 1911 it opened an assembly line in Paris but in the early years it was the Spanish design that dominated. From 1919 the French one led the way with some of the finest cars of the decade but car production ceased in both plants around 1939.

1912 HISPANO-SUIZA ALFONSO Based on a racing Hispano and named after the Spanish king who owned many examples of the marque, this was the first great car from the firm. It was one of the earliest true sports cars and matched some much larger with its modest 3.6-litre, four-cylinder engine.

ABOVE: **1924 HISPANO-SUIZA H6C** This series is rated by many as one of the finest of the 1920s. In this version the six-cylinder, overhead-camshaft engine was enlarged to 8-litres but kept its fully-machined crankshaft and alloy block. It was built in both countries, but the French version is preferred, and fitted with all manner of bodies, this one in tulip wood.

BELOW: **1934 HISPANO-SUIZA T68-bis** One of the most magnificent cars ever built; impressive, elegant and sold as a chassis to which the best coachbuilders of France applied themselves. The V-12 engine began at 9.4 litres but was stretched to 11.3 in this version. Both had overhead valves, an alloy block plus heads, and a great deal of power to transmit via the three-speed gear-box while the car was slowed by massive drum brakes, servo-assisted as on the earlier H6 cars.

ABOVE: **1935 HISPANO-SUIZA K6** A less-extravagant model built from 1934 onwards, but still powered by a 5.2-litre, six-cylinder, overhead-valve engine. Not quite so exciting, but the coachbuilders continued to produce many fine creations with the 1935 car leaving the driver out in the weather.

BELOW: **1937 HISPANO-SUIZA K6 5** The 1937 coupé offered yet another line to a fine, French-built car.

HONDA

Japan's largest motorcycle firm turned to cars in 1962 but found sales disappointing in Europe. Technically advanced, this Honda S800's (below) 791cc, twin-cam engine ran to 8000rpm when contemporaries struggled to exceed 5000 and owners had to adapt and use the revs to obtain the undoubted performance.

HOTCHKISS

From 1903 to 1954 this French firm built good touring and luxury cars, usually conventional, even staid at times, but with enough luxury to be hard hit by the postwar French car tax system.

LEFT: **1934 HOTCHKISS CABRIOLET** An example of the firm's products in the 1930s when they used a six-cylinder, overhead-valve engine of 2.6 or 3.5 litres, the smaller resulting in an underpowered car.

1934 HOTCHKISS SALOON
Like the Cabriolet, the saloon possesses French style of the decade but without any flamboyance – essentially a tool for the job of driving.

HUDSON

The name came from a wealthy Detroit store owner who backed the firm from 1909 and they soon built up a good reputation. They sold really well in the 1920s but were hard hit in the Depression years, not helped by a change from a good six-cylinder engine to a smaller eight. They kept going postwar but merged with Nash in 1954.

1938 HUDSON TERRAPLANE Used as a brand name at first but later simply a Hudson model. Powered by a six-cylinder engine listed in two capacities and available in sedan, coupé and fixed-head coupé bodies, it helped to stretch the range.

1950 HUDSON COMMODORE EIGHT For 1948 Hudson had a new unitary body known as the 'Step Down' thanks to its dropped floorpan. Low, strong, safe and stylish, it sold well but lack of finance meant that the firm was stuck with it for too long, unable to restyle or update when needed. Racing successes in the early 1950s failed to help enough to keep them going.

HUMBER

A name first seen on Victorian bicycle, it went on cars in Edwardian times. Taken over by Rootes in 1931, they concentrated on better-class cars into the 1960s but then the marque declined, to disappear in 1976. The Snipe (below, a 1931 model) appeared in 1930 and the name stayed to 1967.

HUPMOBILE

In 1909 Robert Hupp introduced this model which actually undercut Ford's Model T on price and offered a conventional two-speed gearbox as well as magneto ignition. Hupp left the firm the year this 1911 Model 20 (below) was built but the make prospered through to 1928, after which it was all downhill to a 1941 end.

INVICTA

In 1931 (the year of this roadster, below) Donald Healey won the Monte Carlo Rally in an Invicta, proving beyond doubt its ability to cover distance at speed in comfort and without strain. A 4.5-litre, six-cylinder Meadows engine provided the push via a four-speed gearbox. Production was very limited.

ISOTTA FRASCHINI

Dating from 1901, this Italian firm produced large and luxurious cars from its earliest days right through the 1920s, many going to the United States. This market vanished in the Depression, so production of cars virtually ceased except for a few built from spares stock. By the mid-1930s the firm concentrated on aero engines and trucks. An attempted postwar revival attempt came to nothing.

1930 ISOTTA FRASCHINI TIPO 8A Based on the 1919 Tipo 8, this version was built from 1925 using a larger 7370cc, eight-cylinder engine and bigger drum brakes. A typical radiator mascot of the late-1920s (left) captures the spirit of the time before the Wall Street Crash.

JAGUAR

William Lyons started in business building sidecars in Blackpool but then turned to producing stylish car bodies under the name of Swallow. First was the Austin Seven in 1927, followed by a Wolseley Hornet. Next, in 1932, came the first SS, based on a Standard, and through the decade the cars became more SS, less Standard and adopted the Jaguar model name in 1936. Saloons appeared after the war, still using the Standard engine until 1948 when the XK 120 and its twin-cam motor took over. From then on the adverts said it best: 'Grace...Space...Pace.'

RIGHT **1933 JAGUAR SS1**
For its second year the model was improved and listed with the choice of 2-litre or 2.5-litre, side-valve, six-cylinder engines. The underslung frame was made specially by Standard and the detailing, trim and finish were excellent; something that would continue over the years.

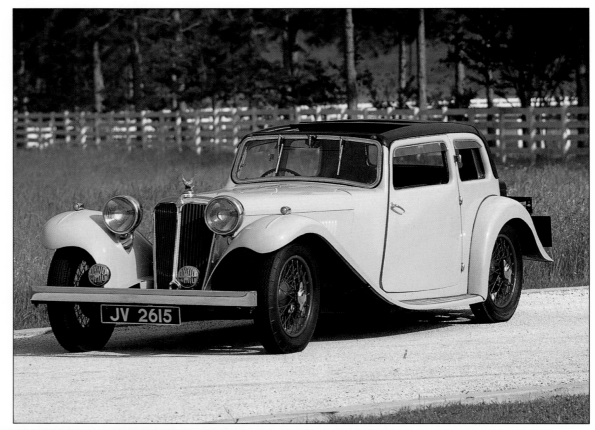

BELOW: **1936 JAGUAR SS100**
The archetypal sports car of the 1930s, mocked by some, but the combination of speed, good style and low price made it a best seller. At first with a 2.6-litre, overhead-valve, six-cylinder, Standard engine, modified for extra power, later with 3.5 litres for greater performance.

1948 JAGUAR 3½-LITRE
Postwar, three engine sizes were available for saloons. The 3½-litre was the largest. These elegant and well appointed models were a lot like the prewar models, and set one of the styles the firm would keep from then on.

1950 JAGUAR XK 120
A sensation when announced in 1948 and now one of the most sought-after classics. The 3442cc, six-cylinder engine with its chain-driven, twin overhead camshafts was to run on for many years in many other models while the XK-series ran for over a decade in several body forms.

1954 JAGUAR D-TYPE Jaguar created the XK 120C to race at Le Mans and won in 1951 and 1953 with the C-type which was also sold to private owners. For 1954-56 they moved on to the D-type which continued with the space frame and disc brakes first used on the C in 1953. The six-cylinder engine pushed out 250bhp and the head fairing was usual.

ABOVE: **1955 JAGUAR MK VII** This regal model was once owned by the Queen Mother, and was modified to the Mk IX standard. Both the Mk VII and the Mk IX retained the separate chassis of the past so the result was large, heavy, luxurious, fast and thirsty. Later cars had disc brakes as well as larger engines.

BELOW: **1957 JAGUAR XK-SS** A rather special sports car of which few were made using spare, monocoque D-type body structures, minus the headrest cum tail fin, but plus interior trim and windows. Originally it came with the dry-sump engine and other D-type features but was often altered to stock XK-parts.

LEFT: 1960 JAGUAR XK 150S
The XK-series moved on via the XK 140 to this final form. The engine was stretched to 3781cc for 1959. The body line lost some of the sculptured early looks although the interior was more comfortable.

BELOW: 1962 JAGUAR E-TYPE
A stunning car that caused the same furore as the XK 120 did in 1948. The well-established 3.8-litre engine pushed it close to 240 km/150 miles per hour, there was all-round independent suspension, disc brakes, a lovely uncluttered body line and the appeal was unchallenged – all for under £1500, plus purchase tax!

**1963 JAGUAR 3.8-LITRE
MK II** The MkI of 1955 was the start of the unitary body, saloon line for Jaguar and led to the MkII in 1959, produced in three engine sizes with better handling and disc brakes. Beloved by TV police series such as *The Sweeny*, it was fast, comfortable and well equipped.

1967 JAGUAR 420 A further development of the older saloons, improved by using items from others in the range. The twin-cam six was enlarged to 4.2 litres. A change of badge turned the car into a Daimler Sovereign, Jaguar having acquired that firm in 1960.

**1973 JAGUAR E-TYPE
SERIES III** In its final form the E-type was given the magnificent V-12 Jaguar engine which more than restored its performance. It was built as a sports two-seater and as a two-plus-two coupé on the longer wheelbase floorpan. It was the last of the series inspired by the racing at Le Mans.

ABOVE: **1987 JAGUAR XJ-S 5.3 CABRIOLET** After the E-type there were no open Jaguars for some years. Removable roof panels adapted the XJ-S coupé body to the cabriolet form, at first with the 3.6-litre six, but here using the 5.3-litre V-12 engine.
BELOW: **1988 JAGUAR SOVEREIGN** The XJ-series dated back to 1968 and ran into the 1990s with a styling update from Pininfarina in 1979. Elegant, finely furnished and powered by the twin-cam six or, as here, the lovely V-12 which gave it pace to go with the grace. Initially listed as XJ12 Series III and as the V12 from 1989.

ABOVE: **1989 JAGUAR XJ-S 5.3 CONVERTIBLE** By stiffening the car structure, Jaguar were able to offer a true convertible complete with an electric-powered hood and able to run to 240 km/150 miles per hour using the V-12 engine. A highly desirable car well able to cruise at 160 km/100 miles per hour with the top down and without ruffling the occupants hair.

BELOW: **1993 JAGUAR XJ 220** The incredibly quick car for the future, as advanced in the 1990s as the XK 120 was in 1948 and the E-type in 1961. Exciting and embodying the best of high technology, the XJ 220 also scored some success for Jaguar at Le Mans.

JEEP

The vehicle that carried the Allies during the Second World War was designed by American Bantam, a firm set up in 1930 to produce Austin Sevens, but most were built by Ford and Willys. The latter, founder in 1908, kept the Jeep in production after the war, but only up to 1955. The name became a marque in the 1960s, to be handled later by American Motors and finally by Chrysler in 1987.

1950 JEEPSTER PHAETON
Still showing its wartime origins but designed as a tourer with a mechanically operated soft top. Four-cylinder and six-cylinder engines, at first with side valves and later with overhead inlets, were listed. The model did well in 1948, its first year, but sales soon dropped off. It was revived for the off-road market and sold in the 1980s, still much the same in layout and looks.

1994 JEEP GRAND CHEROKEE Based on the Cherokee 4.0 Limited, this model's 5.2-litre, V-8 engine made it the most powerful in its class. Four-wheel drive, a high level of comfort, 185 km/116 miles per hour top speed and a full range of luxury interior fitments made a very competitive package.

1994 JEEP WRANGLER This model was available with a 2.5-litre, four-cylinder or 4-litre, six-cylinder engine, manual or automatic transmission and the four-wheel-drive normal for any off-road vehicle. Far removed from the wartime Jeep, it was well fitted out with comforts.

JENSEN

The Jensen brothers were coachbuilders prior to building some nice cars carrying their name from 1936, all having two-speed rear axles and stylish bodies. After the war there were fine grand tourers and later sports saloons using Chrysler V-8 engines, even four-wheel drive as early as 1966. The Interceptor's (below, a 1974 model) name dates from 1949 but continued until 1976 using a 7.2-litre Chrysler engine at the end. Quick but thirsty in coupé or convertible forms. A few more were built in the late 1980s and early 1990s.

LAGONDA

An old British firms, founded by an American, that began with motorcycles before moving on to cars. Between the wars they turned to sports cars and tourers, and in 1947 became part of Aston Martin. This 1934 Lagonda 16/80 (below) is typical of their 1930s range in the lines and style of the period.

LAMBORGHINI

Ferruccio Lamborghini was a tractor manufacturer before he turned to cars in 1963 with the aim to better Ferrari in the supercar league. Stunning looks from some of the top Italian body stylists were matched to an all-alloy, V-12 engine having twin overhead camshafts. After a hesitant start, a line of exotic models followed while the firm went through various hands.

1971 LAMBORGHINI MIURA SV First seen in 1966, five years of development had brought more power from the 3929cc engine set transversely behind the seats in a unitary body. Very quick and very stylish – a model that set the standard for the rest to aim for.

1971 LAMBORGHINI ESPADA In this case the V-12 engine went at the front to enable Bertone to style a body able to seat four on an extended wheelbase. Top speed was down to a modest 250 km/155miles per hour, while the interior was luxurious.

RIGHT AND BELOW: **1982 LAMBORGHINI COUNTACH** The firm had to produce something sensational to take over from the Miura; the Countach, first shown in 1971 and in production in 1974, was it. The V-12, twin-cam engine was in-line mounted in the middle of the car with its gearbox ahead, the drive shaft then running back through the sump to the limited-slip differential at the rear. The assembly went into a tubular frame clothed in an aluminium body having a sharp edge to its wedge style and doors that swung up parallel to the body with help from a servo-mechanism. Later came engine sizes of 4.7 and then 5.1 litres and, in 1985, four-valve heads. In this way the Countach ran on into the 1990s, its 20-year-old shape still a sensation.

1993 LAMBORGHINI DIABLO The successor to the Countach had to maintain the company's reputation of building the ultimate sports car. The engine remained the all-alloy, 48-valve, V-12, but stretched to 5.7 litres and producing close to 500bhp. Naturally, it used the latest of technology and was clothed by a striking body on a tubular chassis built using the latest of exotic materials. In 1993 a Special Edition of 150 cars was built to celebrate 30 years of production, and these had more power and a modified body style. Also new was the Diablo VT which had four-wheel-drive and active suspension.

LANCHESTER

A pioneer firm that was both hidebound in its use of tiller steering to 1911 on some cars and advanced with an earlier patented crankshaft vibration damper used by them and many others. They fitted overhead-camshaft engines and worm rear-axle drive in the 1920s, while this model (below, a 1930 Lanchester Thirty) had eight cylinders and a 4.4-litre capacity. Daimler took them over in 1931.

LANCIA

Established in 1906, this Italian firm has always been known for being technically advanced while retaining the national flair for cars. This continued between the wars and postwar, but financial problems forced a sell-out to Fiat in 1969, after which the flair became stifled.

1929 LANCIA LAMBDA SERIES 8 The Lambda was built for eight years from 1923, all powered by a narrow-angle, V-4 engine having an overhead camshaft. This went into a stressed hull, years before unitary construction became common, and the front suspension was independent. A remarkable combination.

ABOVE: **1953 LANCIA AURIELA** Built through the 1950s, the Auriela was powered by an all-alloy, V-6, overhead-valve engine that began at 1754cc and progressed in stages to 2451cc. This, plus fine suspension, gave it good handling as well as speed although the body was prone to rust.

BELOW: **1990 LANCIA DELTA HF** The Delta, based on and made by Fiat, appeared in 1980 at first with a single-cam engine, then with a 1.6-litre twin-cam. It performed well and was later joined by this turbocharged version which went better while retaining the four-door hatchback body.

LASALLE

A make introduced by General Motors in 1927 to fill a gap in the corporation's range between Buick and Cadillac. The idea worked well for three years. This 1935 LaSalle side-valve, straight-eight (below) was one attempt to retrieve sales but, although the cars were a bargain, the line had to be terminated in 1940. Later attempts to use the LaSalle name in the Cadillac range came to nothing.

LINCOLN

Set up in 1917 by Henry Leland, who had previously established Cadillac, Lincoln was sold to Ford in 1922. For over a decade they built fine luxury cars, often with bodies from America's best coachbuilders, and this continued into the 1930s with the K-series, powered by a 7.3-litre, V-12 engine. In 1936 this was joined by the Zephyr which had radical styling over a smaller V-12 engine and sold in good numbers thanks to a modest price. Its mechanics then formed the basis of the 1940 Continental, a great car conceived by Edsel Ford.

1948 LINCOLN CONTINENTAL Production of the original Continental ended during the war, in 1942, but the model was revived in 1946, still with its V-12 engine, but with a restyled grill in place of the Zephyr one of old. However, the Continental was never a big seller, and production was again ended with the introduction of new designs for the Lincoln range in 1949.

1965 LINCOLN CONTINENTAL The name Continental returned in 1956 as a separate marque, priced high, a loss-leader for Ford but intended to take on the Cadillac market. The MkII became the MkIII in 1958 at a lower price to pull in the customers. The next year the MkIV resumed the name Lincoln Continental, and then ran and ran – from 1961 the only Lincoln model until 1977. It was always the fully equipped car.

LOTUS

Colin Chapman set up Lotus to support his passion for racing and to exploit his many innovative ideas. At first his cars were stark, then aerodynamic, but in either case built for competition. When he turned to road cars with the Elite he was just as innovative and so this continued via Elan, Europa, Eclat and Esprit to today's supercars. Along the way there were many grand prix cars and several world championships.

1970 LOTUS ELAN SERIES 4 The Elite was a coupé having a fibreglass body and a Coventry Climax engine but for the Elan Chapman moved to Lotus-Ford power and a backbone chassis although the body remained a moulding. The series was built to 1974 and was to return in 1990.

1971 LOTUS EUROPA This series retained the backbone chassis but had the engine at the rear and a coupé body. The first cars, built in 1966, used the engine and transmission from a Renault 16, simply reversed to suit the rear-end installation, but during 1971 a switch was made to take the Lotus-Ford twin-cam unit as used here.

1978 LOTUS ESPRIT This successor to the mid-engined Europa reached production in 1976, having stunned the crowds at the 1975 Paris show. The Giorgetto Giugario body sat on the usual backbone chassis and Lotus engine, a turbocharged version was to come in 1980, while the style remains today. This one celebrates a world title and 30 years in the business.

1989 LOTUS EXCEL During 1974 Lotus introduced a new Elite using the backbone chassis and Lotus engine at the front. From this came the Eclat with revised rear body and in 1982 this was replaced by the Excel which used some Toyota parts and ran on in the range to 1992.

1989 LOTUS ESPRIT TURBO
Introduced as the Turbo Esprit in 1980, the name was reversed for 1987 but the 2174cc twin-cam, four-cylinder Lotusengine continued. The car retained the sleek body and pop-up headlamps of old, plus the very considerable performance.

1990 LOTUS ELAN SE This name returned with a Japanese Isuzu engine and, a first for Lotus, front-wheel drive. However, there was also a new raft subframe front suspension system which gave the car unmatched roadholding to go with the power from the 1588cc four-cylinder, 16-valve, twin-cam engine in stock or, as here, turbocharged form. Sadly, the Elan was costly to make so produced no profit for Lotus despite a high price.

1993 LOTUS ESPRIT S4
The year the turbo car gained power steering and a rear spoiler to compliment its high top speed. Small numbers were built for racing with more power and less weight despite the addition of a roll cage.

MARCOS

Jem Marsh and Frank Costain set this firm up in 1959 and began by building cars using marine plywood in their construction. This continued into the late-1960s using a variety of engines including, in this 1968 instance (below), the 3-litre Ford V-6. The firm collapsed in 1972, but Marsh revived it in 1981 using the same formula to run on in the 1990s.

MARMON

Marmon built cars from 1902 to 1933 including some fine V-8s in the 1920s, but in 1931 introduced the V-16 (below) with a variety of body styles. The engine alone was impressive with 8 litres, overhead valves and all-alloy construction to propel most cars to around 160 km/100 miles per hour in silence. An American classic but too expensive to survive for long.

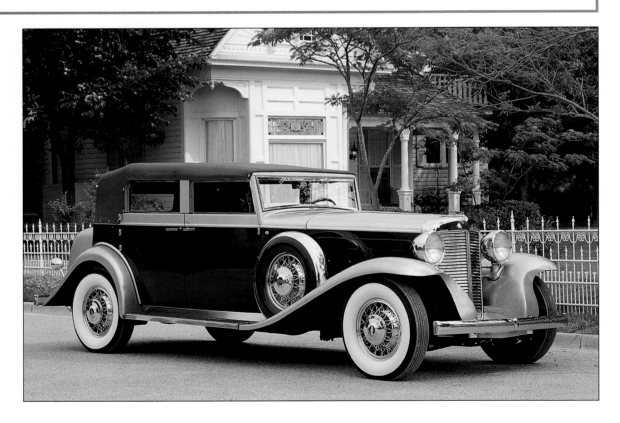

MASERATI

From its beginning in 1926 Maserati built racing cars. Any models sold for the road were merely racers plus lights and mudguards. While a few road cars were built in the late 1940s, they moved from racing to the luxury sporting and grand touring market in 1957. Taken over by Citroën in 1968 and De Tomaso in 1975, the latter firm has revived the Maserati marque's fortunes in recent years.

LEFT: **1993 MASERATI GHIBLI** Typical of Maserati products for the 1990s, the Ghibli used the best materials for its high quality, luxurious interior. Powered by a 2.8-litre, V-6, 24-valve, turbocharged engine producing 280bhp, it could run close to 255 km/160 miles per hour. Among its features was a spoiler at the base of the screen to discourage the wipers from lifting at speed.

BELOW: **1993 MASERATI SPYDER** The Biturbo coupé was designed to be developed into a two-door, two-seat convertible. The coupé appeared in 1984 but it was three years before the Spyder was added, while the V-6 engine's capacity grew from the original 2 litres.

1993 MASERATI SHAMEL
Quick version of the Ghibli with an engine producing some 326bhp, a coupé style and a 270 km/168 miles per hour top speed. The lamps set in the front bumper are for parking, turn signals and to light the way far, far down the road.

MCLAREN

For many years McLaren have been a major player in Formula One. Their F1 road car (below) used advanced technology to reduce weight and drag while raising the power. The result was a V-12 engine, over 600bhp and well over 320 km/200 miles per hour from a three-seat car where the driver sat centrally, in total comfort, to capture fully the Grand Prix experience and feel – at colossal expense.

MERCEDES-BENZ

Formed in 1926 by combining the pioneer Benz firm, the first volume producer, with Mercedes, the first design to break loose from the horse-carriage past, the result was formidable. There was a big range prewar, extending from reliable saloons to aggressive sports cars and luxurious limousines. The firm's grand prix racers were dominant. Postwar, they built a range of quality models to establish the name for advanced technology, excellent finish and reliability. Most were also high performers.

LEFT: **1904 MERCEDES**
Designed by Wilhelm Maybach for Emile Jellinek and named after the latter's daughter, the Mercedes had a pressed-steel frame, honeycomb radiator and gate gear-change. Later cars were designed by Paul Daimler, son of Gottlieb with whom Maybach worked, and then by Ferdinand Porsche.

BELOW: **1927 MERCEDES-BENZ 38/250 SS** Despite the touring style and artillery wheels, this was a very quick car for its time. A Porsche-designed, 7-litre, supercharged, overhead-camshaft, six-cylinder engine provided the 200bhp which pushed it along to well over 160km/100miles per hour.

1928 MERCEDES-BENZ 38/220 S For this, the most sedate of the series, buyers had to be content with only 6.8 litres of the Porsche six engine and 180bhp. Still fast while making gorgeous noises and, for many, having the best looks. Never cheap, even when new.

1936 MERCEDES-BENZ 500K The successor to the S-series arrived in 1934 with a 5-litre, overhead-valve, eight-cylinder engine and, while fast, it lacked the brutal style of the earlier cars. It did offer more comfort and better brakes while the looks were the epitome of the late-1930s Teutonic sports car.

1955 MERCEDES-BENZ 300SL The famed Gullwing coupé body which was powered by a 3-litre, six-cylinder engine installed in the tubular chassis at an angle. It had fuel injection and dry-sump lubrication and was also built as an open roadster with softtop or hardtop.

ABOVE: **1956 MERCEDES-BENZ 190SL** Although this car looked much the same as the 300SL it used a stock 1.9-litre, four-cylinder engine which gave it reasonable performance, much easier maintenance and a similar choice of body styles.
BELOW: **1962 MERCEDES-BENZ 300SL** The open roadster version of the 300 model shared mechanics with the Gullwing coupé and changed to servo-assisted disc brakes after 1961. The front light housings also altered but the rest of the body style remained highly distinctive.

**1962 MERCEDES-BENZ
190SL** This 300SL look-alike
was built from 1954 to 1963 in
much larger numbers and sold
well thanks to a much lower
price. It was also more practical
for most buyers while still able to
exceed 160 km/100 miles per
hour thanks to the overhead-
camshaft engine.

**1966 MERCEDES-BENZ
230SL** During the 1960s the firm
listed ranges of cars to offer a
variety of engine sizes, finish
standards and body types. In this
case the 230 indicates a 2.3-litre,
overhead-camshaft, six-cylinder
engine and SL the sporting,
two-door body. It also had a
good option list and the ability to
cruise at 160 km/100 miles per
hour in comfort.

**1986 MERCEDES-BENZ
300SEL** Continuing an
established theme, the S-class
became better still in the 1980s
using a variety of engine sizes
and configurations. This one had
the 3-litre, six-cylinder engine
and an extended wheelbase for
more rear-seat room, but the
proportions remained just right,
as did the performance.

ABOVE: **1987 MERCEDES-BENZ 500SEC** Based on the S-series but using a 5-litre, V-8 engine and a fine two-door coupé body, this was one of the firm's finest cars, able to cruise the *autobahn* without effort.

LEFT: **1990 MERCEDES-BENZ SL300** While the 1990 SL300 still used a 3-litre, 6-cylinder plus overhead camshaft engine, little else remained the same from the 1950s versions except for the good looks, style and lines. Twin-camshafts, 24 valves, anti-lock brakes, powered hood and all the rest of the technology contained in a fine package behind the three-pointed star.

MERCURY

A Ford division set up in 1939 to plug a gap between Ford and Lincoln, Mercury followed the ups and downs of the industry through the years. This 1957 Mercury Turnpike Cruiser (below) was previewed by a 1956 show car, one of a new line for 1957 and dropped the year after. Built in sedan, coupé and convertible forms, it used the larger of the V-8 engines then fitted.

MESSERSCHMITT

The aircraft firm responsible for some of Germany's best fighters of the Second World War produced a remarkable microcar for a decade from 1953. It had three wheels with two at the front, tandem seating raised by a hand lever when entering or leaving, a hinged canopy, small two-stroke engine which ran backwards to provide four reverse gears, easy parking, and an exciting ride.

1955 MESSERSCHMITT KR200 Early models had a 175cc engine unit which lacked reverse and had to be kick-started, but the later KR200 enjoyed a full 192cc and electric start. The air-cooled, single-cylinder Sachs engine went at the back, while suspension was by rubber, the brakes cable operated and steering by handlebar.

1960 MESSERSCHMITT TG500 TIGER In this form the model had an extra rear wheel and a 493cc twin-cylinder engine so was good for some 120 km/75 miles per hour. The make differed from the other bubblecars of the period and the layout made it a great vehicle for threading through the traffic jams of its time.

MG

Cecil Kimber created these two magic letters from Morris Garages which he managed and attached them to some sporting Morris models. This enterprise, supported by William Morris, had become Britain's most successful producer of small sports cars at a reasonable price. This continued postwar under BMC (British Motor Corporation), but with much less success as British Leyland when MG became just a group badge.

1930 MG MIDGET The first of a long line, the M-type was introduced in 1929 based on the Morris Minor. Both used an 847cc, four-cylinder engine with an overhead camshaft driven by bevel gears and a vertical shaft that carried the dynamo armature. A lowered Morris chassis and sports or coupé body gave a cheap 105 km/65 miles per hour and an option of four speeds for 1930.

1934 MG NB MAGNETTE By
adding two cylinders, as used in
the Wolseley Hornet, the Midget
four became the Magnette six,
extended to 1271cc for this series
but retaining the overhead
camshaft. With four speeds and
130 km/80 miles per hour, it was
available open with two or four
seats, or as a coupé.

1935 MG NB MAGNETTE
This is the Airline coupé version
of the N-series Magnette which
used the same mechanics but
kept its passengers in the dry.
Very much in the charming
British style of the decade.

1935 PA MIDGET At this point
the 847cc engine was given a
third main bearing, installed in a
longer chassis and there were
other changes, but not to the
basics of sporting motoring and
cutaway doors. The PB version
had a larger engine, but after that
model the camshaft was replaced
by pushrods.

1937 MG VA This was more a tourer than other models, built with saloon, coupé or open bodies and powered by a 1548cc, four-cylinder, overhead-valve engine. It was stylish, elegant, well appointed and able to better 130 km/80 miles per hour and stop when needed to do so thanks to large Lockheed hydraulic brakes. This body is by Tickford.

1950 MG TD MIDGET The postwar Midget continued in 1250cc TC form, then developed into this type with independent front suspension, rack and pinion steering and the less-popular disc wheels.

1951 MG TD MIDGET Wire wheels returned on the TF as an option to take the Midget on to 1955, after which came the modern sports MGs.

1955 MG MAGNETTE ZA A combination of the 1489cc four-cylinder, overhead-valve Austin engine and a Wolseley 4/44 body shell produced this model which handled better than an Austin and had a larger engine than the Wolseley. The result was pleasing and performed well.

1958 MGA 1500 ROADSTER The modern replacement sports MG introduced in 1955 using a 1489cc BMC B-type engine having twin carburettors and able to push along to 155 km/95 miles per hour. Separate chassis, independent front suspension, good drum brakes and the option of wire wheels completed the package. The model was also built as a coupé, and could be fitted with a twin-cam or a larger engine until replaced by the best-selling B-type in 1962.

1973 MG MIDGET 1275 The Midget name returned in 1961 on what was also the Austin-Healey Sprite, both using the 948cc and then the 1098cc BMC engine. They moved on to 1275cc in 1966 and continued up to 1979 but in this MkIII form were good for 155 km/95 miles per hour and kept the original Midget concept of cheap motoring fun alive.

ABOVE: 1980 MG B ROADSTER
A very popular model, especially in MGB GT form, but one that gradually fell behind under the weight of black safety bumpers, emission gear, the old engine design and a steady increase in cost. Earlier models are preferred for their looks and sparkle.

LEFT: 1994 MG RV8 The MG name only lived on in the 1980s as a badge on various group cars, but returned in 1992 as a classic reissue. Powered by the 3.9-litre Rover V-8 engine it sounded right and went quickly in a straight line but retained the handling and ride of the 1970s. Time had moved on so the model was an expensive anachronism.

MITSUBISHI

This huge Japanese corporation first built cars in 1917 but only for five years. After the war they produced scooters, then trucks and, in 1960, a small economy car. From this small automotive beginning they grew and grew, selling as the Colt, actually a model name, and then reverting to Mitsubishi in the 1980s. There were many cars and many models over the years, but in time the range was simplified.

1987 MITSUBISHI STARION TURBO A quick coupé powered by a turbocharged 2.0-litre engine which suffered from turbo lag. Good handling but with a firm ride, the car moved on to a larger engine to cope with emission-control gear.

1995 MITSUBISHI SHOGUN This tough vehicle, built for the off-road market, remained pleasant to drive on the road. Four-wheel drive and various petrol or diesel engine options in a 1995 line-up comprising six models and 11 variants, including three engines new for that year.

MORGAN

This firm has achieved a huge success by remaining in the past. Still in great demand, today's cars continue to use the same independent front suspension technology that was fitted to the first Morgan three-wheeler of 1910. Three-wheeled Morgans were built up to 1952, and were joined by four in 1935. They continue totally conservative – but therein lies their charm.

LEFT: **1946 MORGAN 4/4** No real change from 1935 other than to a Standard 10 engine built with overhead valves and a Moss gearbox. The brakes continued mechanical while the rest of the car and body was as prewar. It was not to change much over the next half century.

BELOW: **1990 MORGAN PLUS 8** Little changed in chassis or body so the hard ride and prewar handling remained, but this version went faster thanks to the 3.9-litre Rover V-8 engine crammed under the bonnet. A violent, white-knuckle – yet popular – ride.

MORRIS

William Morris began with bicycles, moved on to a few motorcycles, but the move to cars in 1912 took him to the position of Britain's largest car builder. He achieved this through sound specifications, good quality materials and cost-conscious component purchasing. But from holding half the market at the end of the 1920s, the company slipped, recovered then merged with Austin in 1952. The Morris name was finally dropped in the 1980s.

RIGHT: **1924 MORRIS COWLEY** The famous Bullnose was built up to 1926 in Cowley and Oxford forms, the latter having superior fittings. It became Britain's most popular car thanks to its inherent strengths, but when it became the Flatnose in the late 1920s, it lost its highly distinctive form.

BELOW: **1935 MORRIS EIGHT** The Morris line became crowed and confused in the early 1930s, but 1935 saw the start of a better time. The Eight was the smallest and most successful, powered by a 918cc, side-valve, four-cylinder engine driving a three-speed gearbox. Both saloon and touring bodies were listed, this open model having four seats rather than just the two.

1949 MORRIS MINOR The name dated back to 1929 but this was the start of a long-running model which stayed to 1971 and is now being rebuilt as new for those who love them. The early cars had to manage with 918cc and side valves but all handled well and they became faster with larger engines over the years.

1964 MORRIS MINI COOPER S Both Austin and Morris versions of the Mini were built from the start in 1959 when the classic design first appeared. The Cooper and then the Cooper S were the performance models in the years the Mini won the Monte Carlo Rally and carved up Jaguars at Brands Hatch and Silverstone.

NAPIER

A famous British precision-engineering firm that had long been in business when they built their first car in 1900. They introduced a six-cylinder engine in 1903, a world first for a production car, set a 24-hour world record at Brooklands in 1907 (below, a 1907 Napier) and built aero engines during the war. From 1919 to 1924 they built just one model, the 40/50, to compete with the best but production stopped after 187 cars.

OLDSMOBILE

Ransom E. Olds was an American pioneer who built cars as early as 1896 and was in series production by 1903. Oldsmobile became part of General Motors in 1908. Between the wars they had a reputation for solid and reliable cars. During the 1930s they led the way towards automatic transmission. In postwar years they were a major player in the General Motors conglomerate, often the most innovative member.

RIGHT: **1953 OLDSMOBILE 98 SEDAN** This stock sedan made up the bulk of sales and was listed in two wheelbase lengths, this the longer. Both used the same V-8 engine while the basic body style ran from 1950 for much of the decade.

BELOW: **1970 OLDSMOBILE CUTLASS** This model range was one of the most popular during the 1970s and was offered in a large selection of body styles and two wheelbases, the shorter for the two-door types such as this coupé. There was also a range of engine options, one a six-cylinder, but most V-8s in two capacities and several degrees of tune.

OPEL

This German bicycle and sewing machine firm built its first car in 1898 and went into series production in 1902. It became part of General Motors in 1928 and was a major German manufacturer between the wars. After 1945, Opel became the dominant GM force in Europe. This 1912 Opel 5/14 PS (below) highlights the good basic features which brought the firm its success.

PACKARD

In 1899 James Packard set out to build a better car than the Winton he had purchased. His efforts led to the first Packard six-cylinder engine, which made Packard's name as a prestige car builder. In 1915 Packard brought out the Twin Six, America's first production V-12, and in 1923 turned to a straight-eight engine and continued to dominate the luxury market. During the Depression the firm sensibly moved into the middle-price range. After the war, Packard lost the luxury market to Cadillac. While innovative technology was produced around 1955, following the merger with Studebaker in 1954, the name vanished in 1958.

1932 PACKARD 840 The straight-eight engine still had another decade to serve but in this case the 6318cc side-valve went into this Custom Super 8 model, one of several body styles listed, all with the distinctive ox-yoke radiator and bonnet line.

1938 PACKARD SUPER 8 A *V* windscreen went on some of the cars for 1938 while this Super 8 had the 5342cc version of the venerable but still-effective engine. A fine car in the tradition of the old-style prestigious Packard but the last year the series had its own bodies.

1954 PACKARD CARIBBEAN Introduced in 1953 as a limited edition convertible with 750 built, the model sold well so another 400 were built for the next year. For 1955 they were restyled, given a highly innovative suspension system and the old straight-eight engines were changed to much improved V-8s.

1956 PACKARD CLIPPER This line was added in 1953 at the bottom end of the Packard range. In 1956 Clipper was listed as a separate make before returning to the Packard fold for the marque's final two years. V-8 powered and offered in several well-trimmed and styled forms, this is the rare hardtop.

1958 PACKARD HAWK
Simply a more luxurious Studebaker Golden Hawk given the Packard's name in the marque's last year. Its massive fibreglass nose and inelegant tail fins were far removed from the dignified models associated with Packard's past.

PIERCE ARROW

Based in Buffalo, the firm's first car appeared in 1902 as a Pierce, using models names of Arrow and Great Arrow, hence the change in marque name in 1908. A six-cylinder model came in 1906. In 1913 they adopted their fender-mounted headlamp style and built large, luxurious cars of great style and elegance. They merged with Studebaker in 1928 but became independent again in 1933 to run on to their 1938 end. Before then, they built models such as this 1930 Phaeton (below), able to compare with the best in the world of that time.

PLYMOUTH

Set up by Chrysler in 1928 for the budget end of the market, this marque was a success from the start. It did well in the 1930s by offering style, technical advances and competitive prices. This value-for-money policy, allied to sound engineering, good looks and good quality kept the customers coming back year after year.

LEFT: **1957 PLYMOUTH FURY** Top of the range hardtop coupé powered by their largest V-8 engine with 290bhp under the hood, options with fuel injection to come the next year. The style was stunning and all-new, having a lower beltline and higher tailfins than the opposition, to give it a three-year lead.

BELOW: **1970 PLYMOUTH SUPERBIRD** Part of the Plymouth Road Runner series and evolved from the Dodge Charger Daytona for the same purpose of long-distance racing, the Superbird shared the long nose, hidden headlamps and high-level wing along with the fruits of victory. It looked fast, ran fast, and in race trim could reach 355 km/220 miles per hour. They produced 1920 of them in that one year.

PONTIAC

Created as a model by General Motors to sell as a low-price Oakland, like the cuckoo it took over the nest. The Oakland range went in 1932 and Pontiac remained a major player. Pontiacs stayed conservative and successful in the 1930s and for some years after the war. In the 1960s a move to the performance market allied to a strong range of standard cars revived a flagging marque. This combination continued on over the years to keep the name high in the sales lists.

LEFT: **1937 PONTIAC CONVERTIBLE** From 1935 Pontiac listed both straight-six and eight-cylinder engines, both enlarged for 1937 to 3649cc and 4079cc respectively, and a whole range of bodies, all of which had the grill revised each year.

BELOW: **1956 PONTIAC STAR CHIEF** This is the Custom Catalina hardtop coupé from the Star Chief group, one of the firm's best sellers and powered by their overhead-valve, V-8 engine introduced in 1955 to replace the old side-valve, straight-eight. The Star Chief was listed from 1954 to 1966.

1966 PONTIAC GTO First seen in 1964, the GTO was based on the Tempest to which was added a bigger and better engine, floor gearshift, modified suspension, steering, brakes and transmission. The result was a car able to carry six and out-accelerate a Ferrari whose owners were less than happy with the use of the GTO name-tag.

1978 PONTIAC TRANS AM Inspired by Firebirds running in the Trans-American races, this series began as an option package in 1969 and became a hot car popular with the young. Performance remained high thanks to continued engine options and the model name became better known than that of Pontiac itself.

1991 PONTIAC TRANS AM GTA Keeping the faith as a performance muscle car, the Trans Am moved on into the 1990s in a modern form and style, still using V-8 power in stock or plus options. A fine car with a long life.

PORSCHE

Dr Ferdinand Porsche worked for several other firms before setting up his own design studio in 1930. The Volkswagen was just one project but this enabled the firm to base the first Porsche car on VW parts, some stock, but more and more to be modified as time passed. Initially built in Austria, they moved to Germany in 1950, having already set their standards on high technology and to build the best of sports cars. First the 356, then the 911 with its incredibly-long production run from 1965, and all the others which kept to this theme of style, performance and being the best.

LEFT: **1955 PORSCHE 356 SPEEDSTER** The 356 was the first production Porsche and was listed with a choice of three engine sizes and coupé or cabriolet bodies, these joined by the Speedster in 1955, a model planned to sell at a low price in the United States, which it did successfully.

BELOW: **1959 PORSCHE 356B SUPER 90** The 356 was updated for 1956, to run until 1965, and soon concentrated on one 1582cc flat-four engine in different states of tune. This one had a 90bhp output, hence its name and list type of 1600S-90.

ABOVE: **1973 PORSCHE 911 CARRERA RS** The 911 Porsche was first seen in 1963, went into production late the following year and is still listed, retaining the same body shape, style and flat-six, overhead-camshaft engine. The Carrera name went back to 1956 and the 356. It was used for competition cars at first, and later applied to these and the quickest road models. This is the first such in the 911 series.

BELOW: **1979 PORSCHE 911 TURBO** The turbocharger was added to the Carrera model in 1975 and the capacity stretched to 3299cc in 1978. The body style was essentially unaltered but has added the flared wheel arches to accommodate the fat tyres, the rear spoiler to keep the tail down, and the front bumper to meet the US rules.

LEFT: **1988 PORSCHE 911 TURBO S** This model has the optional slant-nose package which was listed as the 930S and included retractable headlights and air scoops ahead of the rear wheels to make an impressive motor car.

BELOW: **1988 PORSCHE 911 CARRERA CABRIOLET** The 911 line ran on, year after year, usually with the choice of coupé, cabriolet or Targa bodies, the latter having a detachable roof section but fixed rear window to offer a choice between open or coupé motoring.

RIGHT: **1988 PORSCHE 959**
The fabled supercar, built in small numbers, powered by a very special flat-six engine embodying twin turbochargers, fuel injection, 24 valves, and pushing out some 450bhp. This drove through a six-speed gearbox to all four wheels with over 305 km/190 miles per hour available from this highly expensive package.

BELOW: **1989 PORSCHE 911 SE CABRIOLET** The vents just ahead of the rear wheels were added to aid cooling when the turbocharger was fitted while the rear spoiler, or ducktail, assisted high-speed stability. The headlamps were concealed in a style introduced as an option in 1987.

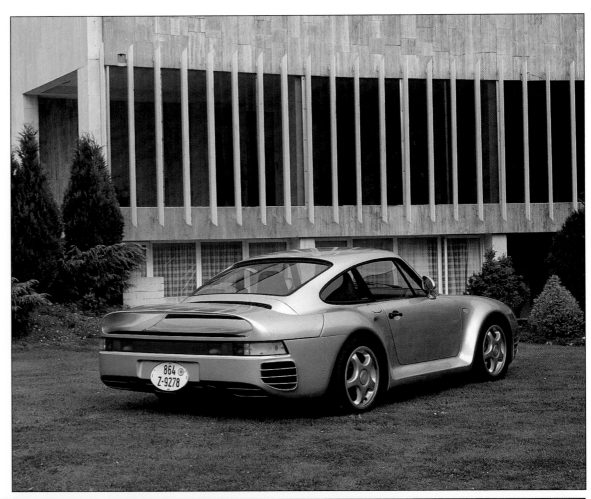

1990 PORSCHE 944 CABRIOLET This popular series continued the theme of a front-mounted, water-cooled engine, as first seen on the 924, and a reversal of usual Porsche thinking. It used a four-cylinder engine with internal balance shafts to smooth the vibrations, the original 2497cc stretching to 2990cc when this cabriolet was listed.

1993 PORSCHE 968 CABRIOLET This model replaced the 944, used its 2990cc engine, but added their Vario-Cam timing system to it. There were six speeds in the gearbox and later came the Club Sport, a lighter version intended for competition.

PROGRESS

The 1901 Progress (below) was a true pioneer showing its horse-carriage origins although steering was by wheel rather than tiller. The occupants sat vis-à-vis – facing one another. The firm used De Dion or Aster engines and its products were renamed the West-Aster in 1904.

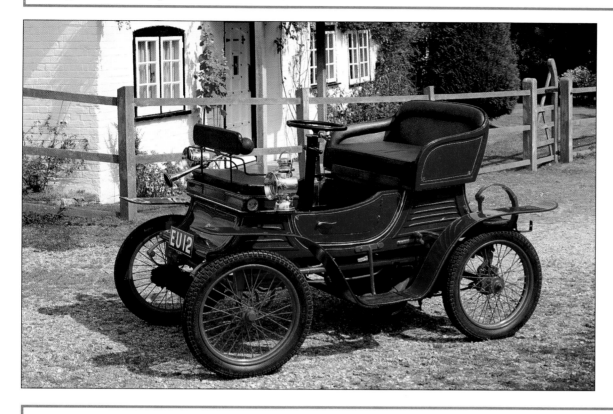

RENAULT

Louis Renault founded this French firm in 1898 and today it remains one of the world's major car producers. Their cars had a highly characteristic bonnet line for years thanks to the rearward radiator mounting. Between the wars they built a good range to cover most needs, but the cars were conservative in design. After the war the state took the company over, and concentrated on cars with rear-mounted engines for a time before moving it to the front. Some were just cars, but others became classics.

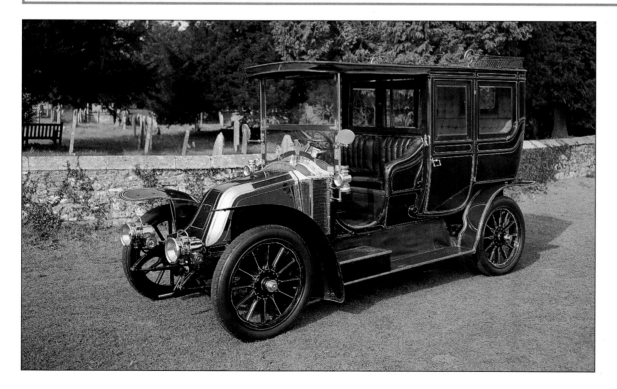

1906 RENAULT The year that Renault won the first French Grand Prix saw this typical model in the true Edwardian style, very much a carriage from the past with the passengers inside and the chauffeur outside, although the latter did have the benefit of a windscreen.

ABOVE: **1958 RENAULT 4CV** Built in large numbers from 1946 to 1961, this model had a rear-mounted, 747cc four-cylinder, overhead-valve engine, rack-and-pinion steering, independent suspension all round and hydraulic brakes, but only three speeds. Good for 95 km/60 miles per hour but a tuned version was built for a few years, this having four or five gears and another 30 km/20 miles per hour.

BELOW: **1972 RENAULT ALPINE A110** Jean Rédelé founded his Alpine firm in 1952 to build sports and competition cars using Renault cars as a base. The A110 was based on the Dauphine but with a much altered engine, chassis, suspension and body. In time his enterprise was absorbed into the Renault group.

1983 RENAULT 5 TURBO
The Renault 5 was a highly-successful hatchback for the 1980s in a classic style with its front-mounted engine driving the front wheels. A great ride, considerable roll angle when cornering and the fullest use of all the inside space added up to timeless charm. However, this one is an aggressive special for road and track with a rear engine, and able to better 190 km/120 miles per hour.

1988 RENAULT GTA A V-6, turbocharged engine pushed this fine sports coupé to over 255 km/160 miles per hour while great handling and grip made the rear-engined car one that encouraged use of its performance. The body accommodated two plus two in comfort while outperforming most others on the road.

1994 RENAULT CLIO WILLIAMS The Clio was Renault's new small car for the 1990s and their association with grand prix racing and the Williams team inevitably led to this limited edition hot hatch model to join the other variants from base RL upwards.

ROLLS-ROYCE

In 1904 the Hon. Charles Rolls and Henry Royce met and agreed to work together to produce and sell a range of cars built to the best standards. This they did and, while their slogan of 'The Best Car in the World' may have been presumptuous at times, they always set the mark that others sought to emulate and better. Inevitably, some do, but none have remained at the pinnacle for so long or in such style.

1907 ROLLS-ROYCE SILVER GHOST Prosaically listed as the 40/50, this model took over from the early Royce cars with two, three, four or six cylinders, and was the only Rolls-Royce built until joined by the 20 in 1922. This particular car was the 13th built and had a silver-painted body, silver-plated fittings and the Silver Ghost nameplate, the first use of what became the model name. It carried out a very-long distance test and firmly established the silent running, refinement and smoothness allied to a fine level of performance characteristic of the Rolls-Royce.

1909 ROLLS-ROYCE SILVER GHOST The Ghost had a six-cylinder, side-valve engine, initially of 7036cc and of 7428cc from 1910. It ran at a modest speed, producing modest power for owners who wanted to drive from walking pace to 115 km/70 miles per hour in top gear without noise or vibration. It had a mechanical form of cruise control, built with the watch-like precision typical of Royce, so the car would hold its speed constant.

LEFT: **1911 ROLLS-ROYCE SILVER GHOST** All manner of bodies went on the Ghost chassis, including a streamlined one which ran at over 160 km/100 miles per hour at Brooklands in 1911, but this one would have concentrated on the delights of motoring rather than speed. It reflects a time soon to disappear for good.

1912 ROLLS-ROYCE SILVER GHOST Every detail of a Rolls-Royce was meticulously designed to be totally functional, long-lasting, efficient, light and easy to make. Royce applied this concept throughout his work and, while he happily adopted others' ideas, inevitably he would improve on them. This applied equally well to an enclosed saloon, such as here, as to an open tourer.

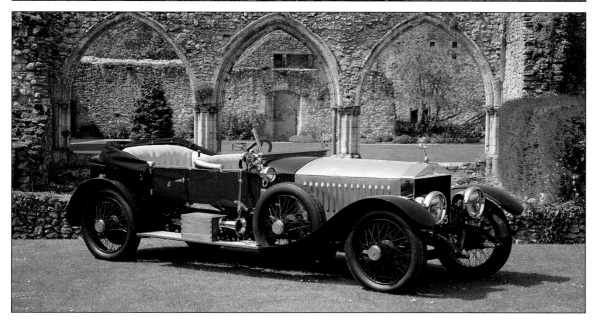

1914 ROLLS-ROYCE CONTINENTAL This model is often referred to as the 'Alpine Eagle' due to the firm's involvement and successes in the Austrian Alpine Trials of 1913 and 1914 which they won. For it, a new four-speed gearbox was used in place of the earlier three speeds and the original four which had a direct-drive third, for silence, and overdrive fourth. The original had been dropped to suit customers requirements for top-gear motoring..

1922 ROLLS-ROYCE SILVER GHOST Nearing the 1925 end of the Ghost but still no front-wheel brakes which arrived in 1924 with a servo system based on those of Renault and Hispano-Suiza, modified and improved to meet Royce's high standards.

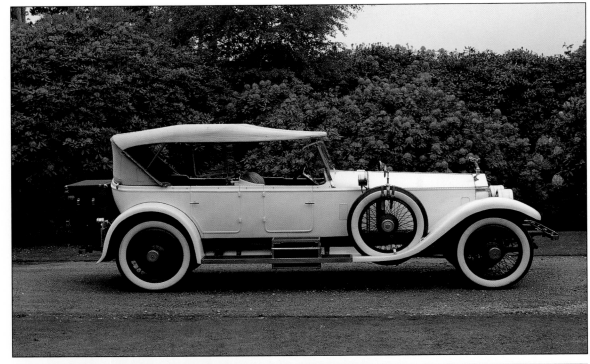

RIGHT AND BELOW: **1931 ROLLS-ROYCE PHANTOM II** The 40/50 New Phantom first appeared in 1925 using a 7668cc, six-cylinder engine with overhead valves. It embodied all the well-developed technology of the Ghost and, as with that model, was also built at Springfield in the United States. In 1929 it was superseded in Britain by the Phantom II which used the existing engine in a new chassis topped by some superb bodies – truly magnificent cars built in the year the firm acquired Bentley.

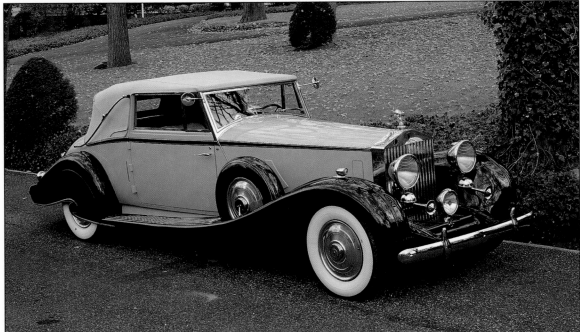

ABOVE: **1936 ROLLS-ROYCE PHANTOM III** Moving on for the late 1930s, the Phantom was fitted with a 7340cc, V-12 engine and independent front suspension while retaining the exceptionally high build standards. In this case, not the most elegant of bodies but no doubt to a customer's tastes and wishes.

BELOW: **1938 ROLLS-ROYCE PHANTOM III** A much better line for a year when the engine had some changes, an overdrive was added and the result was good for 135 km/85 miles per hour in silence and comfort. However, it was a complex car and this made it less popular than it might have been.

ABOVE: **1955 ROLLS-ROYCE SILVER DAWN** After the war, the Silver Wraith became the limousine while the Silver Dawn had a standard steel body. Based on the R-type Bentley, it had a 4566cc six-cylinder engine fitted behind the magnificent radiator and mascot.

BELOW: **1958 ROLLS-ROYCE SILVER CLOUD** This took over from the Silver Dawn for 1956 using much the same engine in a new chassis under the saloon body. Only the radiator and bonnet distinguished it from the Bentley but the cachet of the Rolls name kept the buyers happy.

LEFT: **1963 ROLLS-ROYCE SILVER CLOUD III** For the 1960s the engine became a 6230cc V-8, which had some problems at first. There were four headlights flanking the radiator and, inevitably, a Bentley version of this imposing car.

BELOW: **1963 ROLLS-ROYCE PHANTOM V** The Phantom series ran on postwar, first with a straight-eight engine but with a V-8 for this massive vehicle. Fine proportions masked the 20-feet length while its top speed of 160 km/100 miles per hour would soon empty the 23-gallon petrol tank, a matter of little concern to most owners.

RIGHT: **1974 ROLLS-ROYCE PHANTOM VI** In 1968 this dignified vehicle took over the task of conveying the stately about their business. Mulliner Park Ward built most bodies, technically the Phantom VI lagged behind industry standards, fitting drum brakes to the end, but this was of little moment to the owners from the Queen down. Truly, a majestic car in every sense.

BELOW: **1975 ROLLS-ROYCE CORNICHE** This model took over from the Silver Shadow in 1971 and continued to use the 6.7-litre, V-8 engine. It was built in saloon and convertible forms but was little altered from the older cars in either style or mechanics.

1987 ROLLS-ROYCE SILVER SPIRIT Other cars might seek the 'Best Car in the World' label but continued development and refinement kept the Rolls in contention. Nothing could detract from its long-standing leading position or the fine manner in which it conducted itself. To drive one is to step into another world.

ROVER

Bicycles, motorcycles and then cars were the traditional and much-trodden route in Britain which Rover took, to arrive at four wheels in 1904. Sturdy, competent and reliable in the 1920s, quality middle class in the 1930s, staid and conservative postwar until the advent of the innovative 2000 in 1963, then some bad years of poor quality and design, but finally much success today. Always helped along off-road by the Land Rover and later Range Rover and Discovery.

1937 ROVER 10 The upright style is typical of many Rover models of the decade, although there were sports saloons, and this one was good value for money. It used a 1389cc four-cylinder, overhead-valve engine and sold well, being given a revised body for 1939 to match other later Rovers.

1960 ROVER 100 PA One of a whole series of similar models often referred to as 'Auntie Rovers' by owners. This one used a 2625cc six-cylinder engine with the marque's own form of valve gear, which made it a refined car, well able to better 145 km/90 miles per hour and nicely finished with wood and leather for the interior.

1992 LAND ROVER DISCOVERY Inspired by the Jeep, the original Land Rover remained in production, and was joined by the luxurious Range Rover in 1970. In 1989 the Discovery was added to provide an intermediate slot between the discomfort and basic form of the Land Rover and the expense of the Range Rover.

1993 RANGE ROVER
From 1970 this has led its class, offering an estate car able to go far off-road thanks to its four-wheel-drive and good ground clearance. The V-8 petrol and diesel engines powered it while the level of equipment and finish made it totally civilised.

SINGER

From motorcycles to cars in Edwardian times led to a fine small car in 1912 and a whole series of successful cars in the 1920s to place Singer as third largest in Britain. Too large a range of models caused problems in the next decade, while postwar they fell into Rootes' hands and finished bearing the Hillman badge under Chrysler control. The end came in 1970.

1932 SINGER NINE
Developed from the smaller Junior, the Nine retained its four-cylinder, overhead-camshaft engine but with 972cc and a four-speed gearbox. The saloons, tourers and coupés were nice within their limitations, but the jewel-like engine went on to power sports cars and was built in other sizes and as a six.

1954 SINGER SM1500 Built mainly as a saloon but also in this roadster form powered by a 1497cc engine, still using an overhead camshaft. At least 120 km/75 miles per hour and 10 km/l/30 mpg available in a style of the 1930s and mostly sold abroad.

STANDARD VANGUARD

An Edwardian firm that built good cars between the wars, being especially successful in the 1930s with competent, reliable, conservative and well-priced cars. After the war, they adopted American styling for the tough Vanguard (below, a 1951 model) but then lost this lead to become mediocre and ceased trading in 1963.

STANLEY STEAMER

The Stanley twins were certain that steam was the way to go (below, a 1910 Stanley Steamer) and pursued this dream up to 1917 before selling out, although the firm itself continued until 1927. In their time, they were not alone in their view, but it was soon found that steam was unsuited to the car.

STUDEBAKER

This firm was building covered wagons 50 years before their first electric car of 1902 and the petrol ones that followed. They were soon a major producer, fell on hard times during the Depression, but then recovered to introduce some radical features on some models. They merged with Packard in 1954 but gradually lost out to America's Big Three. The last cars were built in 1966

1931 STUDEBAKER PRESIDENT This name applied to a range of cars built in the 1930s with a straight-eight, side-valve engine of around 5.5-litre capacity at first, later down to 4.1-litre. Quite a classic car with a name to suit, unlike their Dictator range whose name was hardly American or suitable for that time.

1956 STUDEBAKER SKY HAWK This was the hardtop coupé version of the various Hawk models and used the 4.7-litre, V-8 engine. The Hawks had style and performance but their appeal was to enthusiasts so they did little for the volume sales the firm desperately needed.

STUTZ

Henry Stutz ran his first car in the first Indianapolis 500 in 1911 and this was soon followed by the sporting Bearcat. In 1925 the firm set out to build a luxury car offering sporting performance in silence and safety. The result used a straight-eight engine having a single overhead camshaft and was of the best design and quality (below, a 1929 Straight 8). There were many beautiful bodies and in 1931 the engine, already stretched to 5277cc in 1929, added a second camshaft and four-valve heads. The last of these was made in 1934.

TALBOT

A British firm that became entangled with the Sunbeam and French Darracq makes in 1920 which gave rise to a confusing combination and use of the three names. Thus, in the 1930s the Darracq was sold as such in Britain but as Talbot in France. Rootes acquired the British firm in the 1930s, which resulted in improved Hillman and Humber models sold as Sunbeam-Talbots. These reverted to plain Sunbeams in 1954, but the Talbot name returned briefly in the early1980s on Peugeot cars.

1930 TALBOT SIX Georges Roesch designed many fine cars and the 14/45 Talbot first built in 1926 was one of his best. It used a smooth-running, six-cylinder engine of 1665cc which had easily-adjusted, overhead valves, and was well able to be stretched over time and stages to 3377cc. Many body styles were used by this successful series.

1934 TALBOT 105 For this model the six-cylinder engine was enlarged to 2969cc with enough improvements to run the car to over 130 km/80 miles per hour. Again, there were several bodies listed. In Rootes' hands it soon declined and lost its driving appeal even though the bodies kept their looks.

TALBOT-LAGO

In 1935 Anthony Lago joined Talbot to create cars under their joint names. These revived the firm's fortunes on both road and track, but while they continued to win races postwar, their road cars were less successful due, in part, to the high French taxes they were subject to. In 1959 they were absorbed by Simca.

1938 TALBOT-LAGO
A fabulous body by Figoni et Falaschi was typical of the French style of the period applied in place of the stock saloon or coupé. Underneath went a 4-litre, six-cylinder, overhead-valve engine developed to produce plenty of power, while special factory versions won some important races.

1948 TALBOT-LAGO GRAND SPORT The engine was stretched out to 4482cc postwar to produce a *grand routier* motorcar with plenty of power, a Wilson pre-selector gearbox and all the trappings of elegance. As before the war, there were stock coupé and cabriolet bodies, many from French coachbuilders.

TOYOTA

This Japanese firm built its first car in 1935. In the 1960s came massive expansion, exports, assembly plants world wide, and a vast range. The Supra (below, a 1993 model) is a high-performance coupé having a 326bhp engine and able to run to 250 km/155 miles per hour using a six-speed manual or automatic gearbox.

TRIUMPH

After making motorcycles for 20 years, Triumph added cars in 1923 but offered too many models in the 1930s so declined and were bought by Standard in 1944. They built middle-class and sports cars in many shapes and forms in the postwar years, became part of British Leyland and were involved with Rover in the link to Honda, but the name faded in 1981.

1948 TRIUMPH ROADSTER
Introduced in 1946 alongside a similar Saloon, both models were uprated in 1948 to use the Standard Vanguard 2088cc engine and three-speed gearbox, the latter a step back from the earlier four the Roadster had and needed. A classic sports car.

1951 TRIUMPH MAYFLOWER Smaller, 1247cc, two-door saloon which used a prewar, side-valve Standard engine fitted with an alloy head and built for five years to offer dependable motoring with a modest fuel consumption. The razor-edge body style came from the larger cars to which it was better suited.

1964 TRIUMPH TR4 For this model in the TR-series there were many improvements and a restyle which resulted in a longer car. Less stark than the first TR2, with a larger version of the Standard Vanguard engine.

1969 TRIUMPH TR250 For the home-market TR5 and this export model for the United States, Triumph fitted a six-cylinder engine in the TR4A shell. The nose stripe was added to export models, but the engine power went down from 150bhp to 105bhp.

LEFT: **1973 TRIUMPH TR6**
By 1973 the power was down thanks to a milder camshaft. The body style was revised by Karmann and the car did remain fast and tough. The TR7 which followed was a pale shadow of the early cars but the TR8 which came in 1980 used a Rover V-8 engine to get needed punch.

TVR

Set up in 1947, TVR began production of exciting sports cars on something of a kit-car basis. Commercial stability for the Blackpool firm came in the 1960s based on a series using a tubular chassis, firm suspension, fibreglass coupé body and a range of engines. This continued through the 1970s and 1980s, the engines getting larger and the cars faster without losing their character.

1993 TVR CHIMAERA Built in the manner of the old-fashioned sports car that required the driver to drive without any electronic aids such as engine mapping or traction control, although the car did have a needed limited-slip differential. Powered by a 4.3-litre V-8 based on the Rover unit but with twin overhead camshafts. This was a true sports car, rough edges and all.

1993 TVR GRIFFITH Should the Chimaera seem a touch too civilised for the TVR buyer, the firm also listed this car which fitted a 5-litre engine churning out 340bhp. Much else was similar, a basic car, no aids, just even more performance and, thus, a greater need for the driver to display skill in control.

VAUXHALL

Initially based in the London district of the name in 1903, the firm moved to Luton two years later. Soon after, they built their Prince Henry and later the famous 30/98 sporting cars. Late in 1925 General Motors took the firm over and by the 1930s the sporting image had been replaced by more prosaic cars, but ones that sold in volume. This continued in the postwar years, but in time sales fell and the 1970 cars were essentially Opels from Germany, the results uninspiring. However, the 1980s brought much improvement and the firm was once again able to challenge Ford.

1905 VAUXHALL The earliest products from Luton were cars using three-cylinder engines and these models were successful in the trials and hill climbs of the period, but their entry in the 1905 Tourist Trophy failed. These cars were soon replaced by a more conventional design.

1915 VAUXHALL PRINCE HENRY The early classic which did well in the Prince Henry trials in Germany in 1910 and 1911 using a 3-litre engine. It grew to 4 litres for this model and later to 4½ litres for the 30/98 which succeeded it.

ABOVE: **1933 VAUXHALL 14/6 STRATFORD** A sports tourer fitted with six-cylinder, overhead-valve engines of 1530 or 1781cc and produced through the decade in many body styles. Popular, and improved over the years.

BELOW: **1960 VAUXHALL CRESTA** The Cresta name applied to the luxury version of the Velox but had the same mechanics to offer adequate performance, comfort and reliability for buyers who preferred Luton to Dagenham in a style akin to Detroit. The hand of General Motors remained clear.

VOLKSWAGEN

For many years the Beetle was the only car built at the Wolfsburg plant set up in 1945 to produce the design laid down by Dr. Ferdinand Porsche in the 1930s. While it became the world's best seller at way over 20 million built, the firm ran into problems in the 1960s in their efforts to produce a worthy successor. It came in the 1970s as the Golf, while retaining the reputation for production quality and reliability VW had established for itself.

LEFT: **1950 VOLKSWAGEN BEETLE** The first model which had a 1131cc, flat-four, overhead-valve, air-cooled engine located at the rear of the car. Crude, basic, hardly economical, but so reliable and long-lasting that it had enormous appeal for two decades. It continues in production in South America.

BELOW: **1953 VOLKSWAGEN EXPORT BEETLE** This was the version exported into Britain which retained the 1131cc engine but gained hydraulic brakes and synchromesh for the gearbox. Six-volt electrics which were adequate then, and a tap for the petrol tank under the bonnet were further features.

1981 VOLKSWAGEN GOLF GTI The Golf took VW into modern times in 1974 using a transverse, front-mounted, water-cooled engine, front-wheel drive and a hatchback body. The next year brought the GTI version, the first 'hot hatch', and thereby created the major trend of the time and a car that was both practical and fun to drive.

1994 VOLKSWAGEN CORRADO VR6 From its 1989 launch the Corrado hatchback coupé had style, while underneath was an outstanding chassis. The VR6 version came in 1992 using a narrow-angle, 2.9-litre, overhead-camshaft, V-6 engine able to push it along to over 225 km/140 miles per hour to prove how good the car was.

1994 VOLKSWAGEN GOLF VR6 By using the compact V-6 engine, VW were able to accommodate a six in this class of car for the first time. In this application the bore was reduced by 1mm and the capacity to 2.8 litres. In other respects the model was the familiar GTI.

WOLSELEY

Dating from Victorian times, this firm had its first cars designed by Herbert Austin in that era. He left the firm in 1905 and they then built large cars until the 1920s when they turned to overhead-camshaft engines for most models. William Morris acquired the firm in 1927 to introduce new models and the marque became builders of luxury versions of the Morris. This continued in the postwar years, but after the BMC merger of 1952 Wolseley lost its identity and often models carrying the name were identical with other BMC cars except for the badge. Wolseley as a name went in 1975.

1904 WOLSELEY The early cars had horizontal engines, a flat-twin beside the rear wheel for the first three-wheelers, but those with four wheels had a water-cooled single-cylinder engine driving a three-speed and reverse gearbox by chain and thence to the rear axle by a further chain. Two-cylinder and four cylinder engines were also used in this period.

1931 WOLSELEY HORNET The fine overhead-camshaft engine built in the 1920s led to the 847cc four used in the Morris Minor and MG Midget. It then added two cylinders to become the 1271cc Hornet but used an extended Minor chassis which precluded any pretence of handling. However, it proved popular, despite its whippy frame.

1933 WOLSELEY HORNET SPECIAL After three years the six-cylinder engine was shortened by changing the shaft-and-bevels camshaft drive to a chain. The Special had a tuned engine, although the chassis went unaltered, and all bodies were by specialist coachbuilders.

1936 WOLSELEY 14/56 A car using the typical late-1930s and early-postwar body style based on the Morris range but with added touches of luxury. A 1818cc, overhead-valve, six-cylinder engine powered it to 110 km/70 miles per hour which was fine for the time.

1962 WOLSELEY 1500 A Morris by another name, the 1500 combined the good suspension and steering of the Morris 1000 with the group's 1489cc four-cylinder engine. There was also a similar but quicker Riley version of this car.